Molecules: A Very Short Introduction

'a lucid account of the way that chemists see the molecular world . . . enriched with many historical and literary references, and accessible to the reader untrained in chemistry for whom it was written.'
Times Higher Education Supplement

'Ball's writing is sharp . . . and drolly intelligent . . . reliably good and often excellent.'
New Scientist

'In a society of chemical agnostics, it is a brave missionary who tries to reveal its mysteries, but that is what the author has attempted to do – and done remarkably well. . . . At no point does *Stories of the Invisible* sacrifice sound science for sound bites.'
Nature

'Almost no aspect of the exciting advances in molecular research studies at the beginning of the 21st Century has been left untouched and in so doing, Ball has presented an imaginative, personal overview, which is as instructive as it is enjoyable to read.'
Harry Kroto, Chemistry Nobel Laureate 1996

'A must for all those who wish to acquire a basic scientific culture while greatly enjoying it.'
Jean-Marie Lehn, Chemistry Nobel Laureate 1987

'A modern troubadour, [Ball] deftly and happily extols the magic of tiny leprechauns, furiously active in generating energy, assembling machinery, and exchanging fateful ⸻ ⸻ng visible to our

D⸺ ⸺ ⸺ 1986

VERY SHORT INTRODUCTIONS are for anyone wanting a stimulatir and accessible way in to a new subject. They are written by experts, and been published in more than 25 languages worldwide.

The series began in 1995, and now represents a wide variety of topics in history, philosophy, religion, science, and the humanities. Over the next few years it will grow to a library of around 200 volumes – a Very Short Introduction to everything from ancient Egypt and Indian philosophy to conceptual art and cosmology.

Very Short Introductions available now:

Available soon:

For more information visit our web site
www.oup.co.uk/vsi

Philip Ball

MOLECULES

A Very Short Introduction

OXFORD
UNIVERSITY PRESS

OXFORD

UNIVERSITY PRESS

Great Clarendon Street, Oxford O X 2 6 D P

Oxford University Press is a department of the University of Oxford.
It furthers the University's objective of excellence in research, scholarship,
and education by publishing worldwide in

Oxford New York

Auckland Bangkok Buenos Aires Cape Town Chennai
Dar es Salaam Delhi Hong Kong Istanbul Karachi Kolkata
Kuala Lumpur Madrid Melbourne Mexico City Mumbai Nairobi
São Paulo Shanghai Taipei Tokyo Toronto

Oxford is a registered trade mark of Oxford University Press
in the UK and in certain other countries

Published in the United States
by Oxford University Press Inc., New York

First published as Stories of the Invisible 2001
First published as an Oxford University Press Paperback 2002
First published as a Very Short Introduction 2003

British Library Cataloguing in Publication Data

Data available

Library of Congress Cataloging in Publication Data

Data available

ISBN 0-19-285430-5

1 3 5 7 9 10 8 6 4 2

Typeset by RefineCatch Ltd, Bungay, Suffolk
Printed in Spain by Book Print S. L., Barcelona

Contents

Preface

When Alexander Findlay wrote *Chemistry in the Service of Man* in 1916, there was an urgent need to advertise the benefits that chemistry had brought the world. Nine decades later, those writing about chemistry might hope to have been relieved of that burden. But it is not so. In spite of the single most dramatic contribution of chemical art to society – the increase in life span owing to chemotherapeutic health care – Findlay's words still have a familiar ring:

> The people as a whole, being ignorant of science, have mistrusted and looked askance at those who alone could enlarge the scope of their industries and increase the efficiency of their labours.

This same sternness of tone is often not far beneath the surface of efforts today by the chemical industry and its advocates to defend itself against public disdain and censure. One of the problems is that, while the good is taken for granted almost as soon as it is brought to market, the bad sticks in the mind for years. And there is no denying that the attempts by chemicals companies and governments to shirk responsibility for tragedies such as thalidomide and Bhopal, or near-catastrophes such as ozone depletion, have left them with severely diminished credibility to plead their case.

Thus we face the twenty-first century with a pervasive feeling that 'chemical' or 'synthetic' is bad, and 'natural' is good.

The traditional remedy is to list all the good things that chemistry has given us. This list is indeed long, and those who would demonize industrial chemistry probably enjoy many of its products. But I believe that 'chemistry in the service of man' is no longer what we need. For one thing, it perpetuates the impression of a monolithic scientific and technological enterprise universally committed to advancing its own cause. To outsiders, any culture looks monolithic and therefore potentially threatening. It will be a good day when there is more public recognition of how chemists argue furiously with one another about whether this or that product should be banned or restricted, or of the fact that some chemists work in military establishments while others join the blockade outside the gates. Maybe then we will start to see science as a human activity.

But, secondly, chemistry is not simply a thing to be tamed and commandeered into service. It is also what *makes* a man or woman, and the rest of nature too. The negative connotations of 'chemical' and 'synthetic' are hard now to shrug off; but 'molecules' have not yet acquired such colours. And it is by understanding our own molecular nature that we can perhaps begin to appreciate what chemistry has to offer, as well as perceiving why it is that some substances (natural and artificial) poison us and some cure us.

This is why I risk disapproval from some chemists by writing a guide to molecules that focuses to a large extent on the molecules of life – on biochemistry. What I have tried to show is that the molecular processes that govern our own bodies are not so different from those that chemists – I would prefer to say molecular scientists – are seeking to create. Indeed, the boundaries are becoming blurred: we are already using natural molecules in technology, as well as using synthetic molecules to preserve what we deem 'natural'.

In trying to tell these molecular tales, I have benefited greatly from the expert advice of Craig Beeson, Paul Calvert, Joe Howard, Eric Kool, Tom Moore, and Jonathan Scholey, to whom I extend my sincere thanks.

This book began its life as it will end it: as a contribution to OUP's Very Short Introduction series. I am very grateful to Shelley Cox for having sufficient belief in the text to offer it, for a time, an independent life.

Philip Ball

London
January 2001

Acknowledgements

The author and publisher gratefully acknowledge permission to use the following copyright material:

Extract from *The Dalkey Archive* by Flann O'Brien (copyright © The Estate of the Late Brian O'Nolan), reproduced by permission of A. M. Heath & Company Ltd. on behalf of the Estate.

Extract from *The Periodic Table* by Primo Levi, translated by Raymond Rosenthal, copyright © 1984 by Schocken Books, a division of Random House, Inc. Used by permission of Schocken Books, a division of Random House, Inc.

Extract from *Gravity's Rainbow* by Thomas Pynchon, copyright © 1973 by Thomas Pynchon. Reprinted by permission of Melanie Jackson Agency, L.L.C.

List of illustrations

Except where indicated, drawings are by the author

The publisher and the author apologize for any errors or omissions in the above list. If contacted they will be pleased to rectify these at the earliest opportunity.

Chapter 1
Engineers of the invisible: making molecules

The sergeant beckoned the waitress, ordered a barley wine for himself and a small bottle of 'that' for his friend. Then he leaned forward confidentially.

– Did you ever discover or hear tell of mollycules? he asked.

– I did of course.

– Would it surprise or collapse you to know that the Mollycule Theory is at work in the Parish of Dalkey?

– Well . . . yes and no.

– It is doing terrible destruction, he continued, the half of the people is suffering from it, it is worse than the smallpox.

– Could it not be taken in hand by the Dispensary Doctor or the National Teachers, or do you think it is a matter for the head of the family?

– The lock, stock and barrel of it all, he replied almost fiercely, is the County Council.

– It seems a complicated thing all right.

The shortest of short introductions to molecules has already been written, and is far more witty than mine. Flann O'Brien was a man who liked to serve up his erudition over a pint of Guinness, as though he were discussing the potato crop or the terrible state of the roads out of Dublin. We can benefit from some more of the wisdom that Sergeant Fottrell is sharing with Mick in the Metropole Hotel, on Dublin's main street:

– Did you ever study the Mollycule Theory when you were a lad? he asked. Mick said no, not in any detail.

– That is a very serious defalcation and an abstruse exacerbation, he said severely, but I'll tell you the size of it. Everything is composed of small mollycules of itself, and they are flying around in concentric circles and arcs and segments and innumerable various other routes too numerous to mention collectively, never standing still or resting but spinning away and darting hither and thither and back again, all the time on the go. Do you follow me intelligently? Mollycules?

– I think I do.

– They are as lively as twenty punky leprechauns doing a jig on the top of a flat tombstone. Now take a sheep. What is a sheep but only millions of little bits of sheepness whirling around doing intricate convulsions inside the baste.

What is a sheep? This simple question is (under many guises) more than enough to have kept scientists occupied for hundreds of years, and will continue to do so for many years to come. The science of molecules gives an answer embedded in a hierarchy of answers. It is concerned with the 'millions of little bits of sheepness', which are called molecules. A sheep is a blend of many kinds of molecule – tens of thousands of different varieties. Many of them appear not only in sheep but in humans, in the grass, in the skies and oceans.

But science, seeking deeper levels of understanding, does not leave things there. Are not a sheep's molecules made of atoms, and are not atoms made of subatomic particles such as electrons and protons, and are not those made of sub-subatomic particles such as quarks and gluons, and who is to say what *they* contain within their absurdly tiny boundaries?

– Mollycules is a very intricate theorem and can be worked out with algebra but you would want to take it by degrees with rulers and cosines and familiar other instruments and then at the wind-up not believe what you had proved at all. If that happened you would have to go back over it till you got a place where you could believe your own facts and figures as exactly delineated from Hall and Knight's Algebra and then go on again from that particular place till you had the whole pancake properly believed and not have bits of it half-believed or a doubt in your head hurting you like when you lose the stud of your shirt in the middle of the bed.

– Very true, Mick decided to say.

It is indeed an intricate business to work out what molecules are, if you want to begin on a lower (we should perhaps say deeper) rung of the ladder of science and climb upwards. That is necessary if one wishes fully to understand why molecules behave the way they do, and in consequence why matter – why a sheep or a rock or a pane of window glass – displays its characteristic gamut of properties. But many scientists who work with molecules do not need to bother with all the algebra, for its implications can be generally boiled down to rules of thumb about how molecules interact with one another. The chemical industry was a thriving enterprise before chemistry found its mathematics. Which is a way of saying that molecules need not, after all, make your head hurt.

Leaving the table

It is curious that, when Flann O'Brien reworked the conversation between Sergeant Fottrell and Mick from *The Dalkey Archive* into his most famous novel *The Third Policeman*, published after his death in 1966, he systematically replaced the 'Mollycule Theory' with the 'Atomic Theory'. Here then is the very item, the ambiguity about what things are made from. Is it atoms or molecules? Chemists give out mixed messages. Their iconic cryptogram is the Periodic Table, a list of the ninety-two natural

Elements: Primo Levi's *The Periodic Table*

There are the so-called inert gases in the air we breathe. They bear curious Greek names of erudite derivation which mean 'the New', 'the Hidden', 'the Inactive', and 'the Alien'. They are indeed so inert, so satisfied with their condition, that they do not interfere in any chemical reaction, do not combine with any other element, and for precisely this reason have gone undetected for centuries. As late as 1962 a diligent chemist after long and ingenious efforts succeeded in forcing the Alien (xenon) to combine fleetingly with extremely avid and lively fluorine, and the feat seemed so extraordinary that he was given a Nobel prize . . .

Sodium is a degenerated metal: it is indeed a metal only in the chemical significance of the word, certainly not in that of everyday language. It is neither rigid nor elastic; rather it is soft like wax; it is not shiny or, better, it is shiny only if preserved with maniacal care, since otherwise it reacts in a few instants with air, covering itself with an ugly rough rind: with even greater rapidity it reacts with water, in which it floats (a metal that floats!), dancing frenetically and developing hydrogen . . .

I weighed a gram of sugar in the platinum crucible (the apple of our eyes) to incinerate it on the flame: there rose in the lab's polluted air the domestic and childish smell of burnt sugar, but immediately afterward the flame turned livid and there was a much different smell, metallic, garlicky, inorganic, indeed contra-organic: a chemist without a nose is in for trouble. At this point it is hard to make a mistake: filter the solution, acidify it, take the Kipp, let hydrogen sulphide

> **bubble through. And here is the yellow precipitate of sulphide, it is arsenious anhydride – in short, arsenic, the Maculinum, the arsenic of Mithridates and Madame Bovary.**
>
> Primo Levi, *The Periodic Table* (1975)

elements (supplemented by some unstable, artificial ones) arranged in a pattern that helps chemists make sense of them. The most famous book 'about' chemistry is the one that Italian chemist and writer Primo Levi named after this tabulation of matter's building blocks, and it reinforces the impression that chemistry begins with this irregularly shaped grid of symbols. At school I was encouraged to learn mnemonics encoding the elements in the first two rows of the table, which are the most important. For undergraduate chemistry it was required that one could recite the whole thing from memory, to know that iridium lies at the foot of cobalt, that europium is sandwiched between samarium and gadolinium. Yet I doubt that I shall ever set eyes on samarium (although europium shines out at us redly from our television screens).

But chemistry is only incidentally about the properties of the elements, and the science of molecules can afford to ignore many if not most of them. The Periodic Table really belongs to that realm where chemistry becomes physics, where we must wheel out the algebra and the cosines to explain why atoms of the elements form the particular unions called molecules. The table is one of the most beautiful and profound discoveries of the nineteenth century, but, until quantum mechanics was invented by physicists in the twentieth century, one could look upon it only as a mysterious cipher, a kind of crib sheet that served as an empirical reminder that elements come in families whose members show similar proclivities.

Perhaps I am being too quick to dispense with the Periodic Table. At least, I should not do so without confessing to an agenda.

A conventional history of chemistry presents it as a quest to understand matter: to ask, what are things made of? This links chemistry with ancient Greek philosophy, with the attempts of Leucippus and his pupil Democritus to formulate an atomic theory of matter in the fifth and fourth centuries BC. It gives us a narrative that progresses from Empedocles' four elements – earth, air, fire, and water – through to Plato's marriage of elemental theory with atomism (Fig. 1), skirting cautiously around the medieval alchemists' belief in the transmutation of the elements and alighting gingerly on the phlogiston theory of the eighteenth century. We watch Robert Boyle redefine the idea of an element in 1661 (which, however, does not actually amount to much of a redefinition at all), we see antiquity's four-element scheme crumble before the discovery of new 'irreducible substances', and we see Antoine Lavoisier dismantle phlogiston and replace it with oxygen before losing his head under the guillotine's blade in 1794. John Dalton gives us the modern atomic theory in 1800, the list of elements expands enormously throughout that century, and then Dmitri Mendeleev arranges them into the twin-towered edifice of the Periodic Table. The gaps are gradually filled all the way up to uranium (itself known since 1789), and Wolfgang Pauli and the other quantum physicists explain the table's shape in the 1920s.

And so the task is at an end. According to science writer John Horgan in *The End of Science*, this meant that chemistry too was finished, once it had the quantum stamp of approval. The implication in several other recent books on the future of science is that the discipline, conspicuous by its absence, has been consumed from both ends. At the most fundamental level, it has become physics (including that immense but overlooked branch called condensed-matter physics, which ponders on how tangible matter

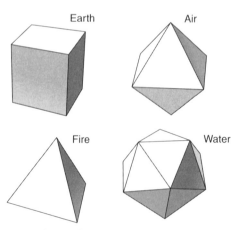

1. Plato's atoms. The Greek philosopher believed that the smallest particles of the four elements then thought to comprise everything had regular geometric shapes.

behaves). At the most complex level, it is now the domain of biologists, who have expanded their world to embrace the molecular mechanics of the cell.

But these academic turf wars conceal a far more interesting truth. It is a curious fact that many histories of science are written by physicists, who have a tendency to present science as a series of questions posed and then answered. It would be instructive to see the story told instead by an engineer, whose instinct might rather be to ask: what can we make? For, while some of our proto-chemists were wishing to dissect matter, whether physically or metaphysically, others were eagerly rearranging it. This is why the science of molecules is both a creative as well as an analytical pursuit. It has been, at various times in history, concerned with making ceramic pots, dyes and pigments, plastics and other synthetic materials, drugs, protective coatings, electronic components, machines the size of a bacterium. 'What is strange', says chemistry Nobel laureate Roald Hoffmann, 'is that chemists should accept the metaphor of discovery'. He goes on:

Chemistry is the science of molecules and their transformations. Some of the molecules are indeed *there*, just waiting to be known by us . . . But so many more molecules of chemistry are made by us, in the laboratory . . . At the heart of [chemistry] is the molecule that is made, either by a natural process or by a human being.

Universities that hide their chemistry departments under the banner of 'molecular sciences' are possibly onto the right idea; for this gently releases the ballast of the Periodic Table and leaves the chemist free to ascend into a world of synthesis, a non-Platonic realm where molecules are designed and made to *do* things, such as cure viral infections or store information or hold bridges together.

As an industrial chemist, Primo Levi moved in this world. He felt a little apologetic about his molecular science: he called it 'a "low" chemistry, almost culinary'. But the power of 'low' chemistry is awesome. It shifts billions of dollars each year, it can make the sick healthy and the healthy sick. Hamburg and Dresden were laid waste by low chemistry, and chemical and biochemical warfare are now more feared in the West than nuclear war. Many people believe that the nuclear bomb was itself the product of physics, but writing $E = mc^2$ does not give you Hiroshima – only separating isotopically distinct molecules of uranium compounds did that. In *Gravity's Rainbow*, Thomas Pynchon has no doubt where the true power of science lies: the villain of his fantasy from the fag-end of the Second World War is not the Bomb but a new plastic, an 'aromatic heterocyclic polymer' called Imipolex G, developed in a conspiracy between Europe's giant chemicals companies IG Farben, Ciba, Geigy, Shell Oil, and ICI. The message is that 'stuff' speaks louder* than theories.

* After the end of the war, a group from the Allies assembled by Eisenhower claimed that 'Without IG [Farben]'s immense productive facilities, its far-reaching research, varied technical experience and overall concentration of economic power, Germany would not have been in the position to start its aggressive war in September 1939.' It was one of IG Farben's subsidiary companies, Degesch, that made the poison gas Zyklon B used in the concentration camps.

Does this mean that molecular science is bad? Of course not – it means that it is a craft full of possibilities. Wonderful, inspiring, inventive possibilities. Terrible, nightmarish possibilities. Mundane but useful things, bizarre things, hard-to-understand things. Molecular science might one day help people to grow a new liver. Raphael, Rubens, and Renoir painted with molecules. Molecules orchestrated the origin of life.

Synthesis: Thomas Pynchon's *Gravity's Rainbow*

The origins of Imipolex G are traceable back to early research done at du Pont. Plasticity has its grand traditions and main stream, which happens to flow by way of du Pont and their famous employee Carothers, known as the Great Synthesist. His classic study of large molecules spanned the decade of the twenties and brought us directly to nylon, which is not only a delight to the fetishist and a convenience to the armed insurgent, but was also, at the time and well within the System, an announcement of Plasticity's central canon: that chemists were no longer to be at the mercy of Nature. They could decide now what properties they wanted a molecule to have, and then go ahead and build it … A desired monomer of high molecular weight could be synthesized to order, bent into its heterocyclic ring, clasped, and strung in a chain along with the more 'natural' benzene or aromatic rings. Such chains would be known as 'aromatic heterocyclic polymers'. One hypothetical chain that Jamf came up with, just before the war, was later modified into Imipolex G.

Thomas Pynchon, *Gravity's Rainbow* (1973)

What are molecules?

So molecules make up everything there is? Not exactly. All matter (outside of some strange astrophysical environments) is made up of atoms; but atoms do not always organize themselves into molecules. (I cannot tell whether Flann O'Brien made the switch from 'mollycules' to atoms because he understood, or did not understand, this distinction.) Most atoms on their own are highly reactive – they have a predisposition to join up with other atoms. Molecules are collectives of atoms, firmly welded together into assemblies that may contain anything up to many millions of them.

But there is a further, subtle distinction to be made. Flann O'Brien's Sergeant Fottrell speaks of 'mollycules' of rock and of iron. Strictly speaking, there are no such things – at least, not in a block of everyday rock or iron. By molecules, we generally mean assemblies of a discrete, countable number of atoms. In the water molecule there are three atoms: two of hydrogen and one of oxygen. A glass of water contains trillions upon trillions of atoms, but a snapshot of the liquid – were it able to reveal such tiny details – would show that at any instant they are nearly all grouped into these three-atom molecules, like a gigantic crowd holding hands in families of three (Fig. 2a).

The atoms in iron, in contrast, do not cluster into discrete molecules. They stack together like cannonballs in a regular array that goes on and on, like a regimented battalion of soldiers. One cannot identify any grouping of the atoms – each is equidistant from its neighbours. The same is true of sodium and chlorine atoms in a crystal of sodium chloride (table salt (Fig. 2b)). When iron melts, the atoms simply jostle one another like an unruly crowd. But when ice melts, it is as if the hydrogen and oxygen atoms continue to hold hands in threes as the crystal falls apart. One would say that ice is a molecular solid – the atoms are clustered into molecules – whereas iron and rock salt are not.

Oxygen   Hydrogen
atom   atom

2. Water (*a*) is composed of discrete three-atom molecules, joined by strong chemical bonds. Salt (*b*), in contrast, is an assembly of charged atoms (ions) of sodium and chlorine, in which there are no discrete atomic groupings. When salt dissolves in water, the assembly merely falls apart ion by ion.

Some pure elements adopt molecular forms; others do not. As a rough rule of thumb, metals are non-molecular, like iron, whereas non-metals are molecular. Frozen nitrogen, for instance, consists of molecules containing two atoms each. In phosphorus the atoms form groups of four; in sulphur they can link into molecular rings of eight. It seems a little unfair that there is no way of knowing, simply by looking at a material, if its essential building blocks are atoms or molecular unions of atoms. But there is not. (It is not hard for scientists to find out, however.)

So 'molecule' is actually a rather fluid, loosely defined concept – essentially a question of scale. Why bother, then, to single out molecules at all, rather than simply talking about 'matter' in general? I would suggest the following reason: molecules are the smallest units of *meaning* in chemistry. It is through molecules, not atoms, that one can tell stories in the sub-microscopic world. They are the words; atoms are just the letters. Of course, sometimes a single letter constitutes a word. But most words are distinct aggregates of several letters arranged in a particular order. We often find that longer words convey subtler and more finely nuanced meanings. And in molecules, as in words, the order in which the component parts are put together matters: 'save' and 'vase' do not mean the same thing.

Some of the most wondrous stories told by molecules take place in living organisms. But unfortunately they can be very difficult to understand: many of the words are long and unfamiliar, and we have only a dim grasp of the syntax. Chemists are constantly inventing new molecular words, expanding the language – and some of these neologisms are rather witty. Some let us tell tales that could not even be formulated before the 'word' was invented. In other cases, a new 'word' allows us to say in a simple manner something that was previously conveyed in a roundabout way.

It is remarkable how nicely the linguistic metaphor fits the molecular world. We hear much today about the 'language of the genes', and I hope to show that this is just one of the tongues that molecules encode. Yet it is not merely a metaphor. There really is 'information' in molecules, just as there is in words, as I show in Chapter 7.

Moreover, using an information-based paradigm to describe molecular science is valuable in so far as it invites a responsive, dialogue-based description rather than the mechanical one that has been championed in former times. Cell biologists speak increasingly about protein molecules that 'talk to' one another; physicists interested in the science of matter speak of 'cooperative' and 'collective' behaviour. These are not woolly, romantic notions calculated to make science appear friendlier (although it will do no harm if they have that effect). Rather, they speak of the increasing awareness of the beautiful sophistication of molecular behaviour, which is generally gregarious and rarely linear.

It is with these thoughts in mind that I need to expand on the use of metaphor in molecular science. We cannot do without it, even at the level of one specialist speaking to another. This is true in many areas of science, but in chemistry more than most. Molecules are anthropomorphized mercilessly, and there need be no apology for that. They are unfamiliar things, these molecules, and we need to find ways of making them less so. The publishers of my book about water rightly insisted that ball-and-stick models of H_2O molecules were anathema to the non-chemist reader, guaranteed to ensure that the book stays on the shelf. Yet I could not explain water's strangeness without showing its molecular structure, and so I made the molecules into little demons (Fig. 3).

I hope this was harmless. But I was reminded of the dangers at a public lecture I attended recently on molecular replication. The first question from the floor was 'Are these molecules conscious?' Given that the speaker was talking about a synthetic molecular system that mimics (in a very crude way) some of the characteristics of living organisms, I suppose this was an understandable enquiry. I firmly believe the answer is 'no', if one wants to retain any meaningful working definition of the slippery concept of consciousness. But, once we start to anthropomorphize, we import a baggage of associations, for better or worse. Many people hate the concept of 'selfish genes' because it carries moral connotations. (Richard Dawkins calls it 'poetic science', and I can see what he means – but the poetry of the mechanism gets besmirched by the unpleasantness, as many see it, of the metaphor.) The idea that molecules 'cooperate' and 'communicate' is no basis for a philosophy of nature. But it is reason to suspect that, in molecular science at least, a linear, clockwork world view might in the end leave us like the ancient astronomers interpreting planetary motions from a geocentric perspective: trying to shoehorn the observations into a misconceived framework.

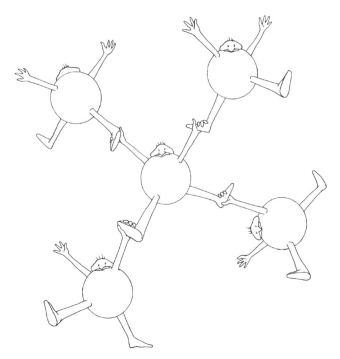

3. Making molecules anthropomorphic can help us visualize how they interact. Here I show the weak 'handclasps' that exist between water molecules.

Shape and size

Primo Levi's *The Monkey's Wrench* is one of the few novels I can think of that includes a drawing of a molecule (Fig. 4). It is a fearsomely complicated one, and I would never dream of showing it in a non-technical book about science if my intention was to be instructive.

But Levi gets away with it, because he does not want us to understand anything about the molecule, except for one thing: it

$$
\begin{array}{c}
\text{I I}\\
\text{N--CO}\\
\diagup \qquad \diagdown\\
\text{CH}_2 \qquad \text{N--}\\
\qquad \diagdown \diagup\\
\text{N--CH}_2\\
\diagup\\
\text{CH}_2\text{--NHCO}
\end{array}
$$

CO—N
−N   CH₂
H₂C −N
            CO—NHCH₂—N
                              CH₂
                  H₂C −N
                              CO—NHCH₂—N
                                                CO—N
                                                      CH₂
                                          H₂C −N
                                                      CO—NHCH₂—

4. Primo Levi's molecule.

has a shape and structure. There are some kinds of hexagon in here, and some straight units linking them together. The narrator is talking to a construction worker named Faussone, a man who assembles girders into bridges. He says,

> the profession I studied in school and that has kept me alive so far is the profession of a chemist. I don't know if you have a clear idea of it, but it's a bit like yours; only we rig and dismantle very tiny constructions ... I've always been a rigger-chemist, one of those who make syntheses, who build structures to order, in other words.

We will encounter in these pages examples of molecules that can be regarded as miniature sculptures, containers, soccer balls, threads, rings, levers, and hooks, all made by sticking atoms together. Plato believed that atoms have the shapes of 'regular polyhedra': cubes, tetrahedrons, octahedrons, and so on. He was wrong;* but chemists can arrange atoms into molecules with these shapes.

---

* Actually, one can make a case that Plato was not far wrong at all. Atoms do link together in quite precise geometrical arrangements. Carbon atoms, for example, like to sit at the centre of a tetrahedron with four other atoms at the corners. This does not exactly make it the tetrahedral block that Plato envisaged for atoms of 'fire'; but it shows that Plato's geometric view of the microscopic world held a grain of truth.

So how big is this molecule that Levi's narrator draws for Faussone? Each one of those C's, N's, and so forth represents an atom, which is a truly tiny thing. Countless analogies struggle to convey the scale of atoms, but I am not sure that they serve to give an impression any more concrete than that these irreducible particles of the elements are very, very small indeed. For example, if a golf ball were blown up to the size of the Earth, its atoms would be about the size of the original golf gall. Ten million atoms of carbon side by side would make a row about a millimetre long.

A small molecule like water is just a few atoms' width in size, about three-tenths of a nanometre. (A nanometre is a millionth of a millimetre.) Primo Levi's molecule is several times bigger. (One cannot say exactly how many times, because what he drew was really just a fragment of a molecule, which continues to the right and the left of the page.)

One consequence of this scale is that things happen very fast in the molecular world. When we hear that molecules can rotate ten billion times a second, we imagine that they must be spinning at unimaginable speeds. But molecules are so small that, even if they travel at quite moderate speeds, they can cover molecular-scale distances in an instant. The atoms of an oxygen molecule need move only at a speed of about a metre per second to complete ten billion revolutions in a second.

What about the sticks that join the atoms together? In fact, they take up no space; they are just a convention to help us see what is going on in the diagram. Atoms that are bound together in molecules push right up against one another; in fact, they overlap, rather like two soap bubbles in contact. This is possible because atoms are not like hard billiard balls, but more like rubber balls. They have a centre that *is* dense and hard, called the nucleus, and this is about ten thousand times smaller than the atom itself – although it is where nearly all of the atom's mass is concentrated. The nucleus has a positive electrical charge. Surrounding it is a cloud of electrons, which are small, light subatomic particles with a negative charge. The electron clouds of two atoms can overlap without danger of electrons colliding, and the two atoms then share some of their electrons: the two clouds merge into one, encompassing both nuclei. When this happens, the two atoms are said to be linked by a *covalent bond*. The sticks in the molecular diagram on p. 15 represent covalent bonds, and they are just a way of helping us to see which atoms are bonded to which.

Here is one of the crucial considerations in talking about molecules, and it is one that complicates the whole attempt: there is no 'best' way of drawing them. One might say: well, never mind the schematic diagrams, why not show what they 'really' look like? But that does not help, because there is no way of taking a photograph of a molecule in the same way as we can photograph a cat or a tree. This is not a matter of technical limitations – it is not that we lack a microscope or a camera capable of resolving such small objects. The fact is the mechanics of seeing make it impossible to 'see' a molecule (or an atom, for that matter) 'as it really is'.

The reason is that we see with visible light, which is a wave-like radiation for which the wavelength – the distance between successive crests – varies from about 700 nanometres for red light to 400 nanometres for violet light. In other words, red light fits about 140,000 undulations into a centimetre. This wavelength is hundreds of times larger than a molecule. Roughly speaking, light cannot be focused to a point smaller than its wavelength, which means that objects smaller than that cannot be resolved.* No light-based microscope will ever show us a sharp image of a water molecule.

---

* I'm speaking here of conventional microscopy, where the light is focused by lenses. There are some new optical (light-based) microscopes that surpass this wavelength-limited resolution by getting the light source up close to the sample and shining it through a tiny aperture. This can increase the resolution to, so far, around a tenth of a wavelength.

I suspect that this is one reason why people find molecules hard to comprehend, and why diagrams like the one above are a good way to scare readers away from a science book. It seems absurd to be talking in a concrete manner about objects that are not only too tiny to see in practice but too tiny to see in principle. Things that cannot be seen acquire an aura of fantasy, as though they are just a convenient fiction.

Molecules are not a fiction, however, and we can prove not only that they are there but that they have definite shapes and sizes. Fig. 5 shows some portraits of molecules taken with a special kind of microscope that does not use light to form its images. Beside each snapshot I show a diagram of the molecule's structure. Well before this type of microscope was invented, the molecules were known to possess these structures; but no one had ever seen them directly. The images are pretty blurry – you would not be able to guess the exact shape of the molecules from these portraits alone. But the shapes seen in the microscope do match up in a convincing way with those expected.

How did we already know the shapes of these molecules before the pictures were taken? Some of the corroborating evidence is experimental. Even though molecules are too small to be resolved with visible light, they can be 'seen' with radiation of a wavelength comparable to their own size. Radiation with a wavelength of about a tenth of a nanometre corresponds to X-rays, and, by bouncing X-rays off crystals, it is possible to deduce where their constituent atoms are located. This means that, if a substance can be made in crystalline form, with all its molecules stacked together in an orderly manner, the technique called X-ray crystallography can reveal the structure of the molecules.

In principle we should be able to see individual molecules with an X-ray microscope that focuses X-rays just as we focus light in an optical microscope. In practice it is very hard to focus X-rays,

a   b   c

○ Hydrogen

● Carbon

○ Nitrogen

5. Molecules 'photographed' with the scanning tunnelling microscope, which is capable of resolving them individually. The STM is not (yet?) capable of showing sufficient detail to allow us to interpret the images without prior knowledge, however.

although scientists are on the verge of being able to do so. In the meantime we can get by with the electron microscope, in which a beam of electrons is bounced off the sample and focused to make an image. Electrons can act as waves too, and using electron waves we can construct images of very large molecules such as proteins or DNA. These pictures are not detailed enough to show individual atoms, but they do give an impression of the molecules' overall shape.

Another way to deduce the shapes of molecules is theoretical: it is possible to calculate them. This gets us into the 'algebra' of Flann O'Brien's 'Mollycule Theory', but there is no need to describe it here. It is enough to say that the laws of quantum mechanics* enable us to predict how atoms will form bonds and where they will then sit in relation to one another. There is nothing arbitrary about the way that atoms join together. In particular, atoms of each element have a tendency to form a fixed number of bonds, which is called its valence. Carbon atoms prefer to form four bonds, hydrogen atoms just one. Oxygen atoms form two.

The quantum theory of molecular structure is indeed 'a very intricate theorem', and even the best computers can solve the equations only approximately. But it is currently possible to calculate the structures of medium-sized molecules with a fair degree of confidence. Comparison of the predictions with the structures of molecules found by X-ray crystallography typically shows a good match. Yet there is still no reliable way of predicting the shapes of many of the big molecules found in living cells. In such cases, X-ray crystallography becomes difficult too, both because it is hard to decode the pattern of X-rays scattered from a crystal of these molecules and because in many cases they refuse to form crystals at all. Cells are full of molecules whose shapes we do not know.

* Quantum mechanics is a mathematical description of matter and its behaviours at very small scales, typically atomic dimensions. At this scale, matter can display wave-like properties.

This is a big hindrance to understanding how life's molecules do their jobs, because a molecule's shape is the key to its behaviour. To reverse a designer's motto, function follows form.

Molecular science is therefore a supremely visual science. Chemists have spent over two hundred years developing pictorial languages to describe their craft, with the result that they now have to be polylingual. There are many different ways to depict molecules, each developed to show the particular aspect that the illustrator wants to emphasize. The English chemist John Dalton began in 1800 by drawing molecules as collections of circular symbols depicting atoms, each embellished with shading or markings to identify the element concerned. It was transparent enough, once you knew the code (Fig. 6).

Very well – but not easy for the printer, who had to make up these symbols specially. A neat shorthand was to give each element a one- or two-letter symbol: C for carbon, O for oxygen, Ca for calcium, Fe for iron. (Classically minded even in the nineteenth century, chemists chose to register the metal as *ferrum*; gold and silver, *aurum* and *argentum*, became Au and Ag in the same way. 'Ir' is not iron but the element iridium. But at least the system was *meant* to be self-evident.)

Then carbon monoxide is simply CO. A multiplicity of atoms of the same element is denoted by a subscript, making the hydrogen molecule H_2.

Yet this scheme does not provide a unique inscription for every molecule. Dimethyl ether and ethanol are different substances with

Carbon
monoxide

Hydrogen

6. Dalton's molecules.

different properties, yet both can be represented by the *chemical formula* C_2H_6O. We are back to the lexicological problem: the meaning of a word is determined not only by the letters it contains, but in what order.

So there is a need for a formalism that shows how the atoms are linked together. This is where the sticks, representing bonds, come in. The two versions of C_2H_6O, which are called *isomers* (same components, different order), can be represented in this way:

A further complication is that the molecule occupies, not the two-dimensional plane of a page, but all three dimensions of space.

7. **This pair of stereoscopic images allows molecular shapes to be seen in three dimensions. The molecule shown here is a form of the enzyme lysozyme, which is present in saliva. Coiled (helical) sections of the protein chain are clearly visible here. Place the page about 8 inches from your eyes, and cross your eyes slightly so that you can see three images. Focus attention on the middle one – in a few seconds, it should become sharp.**

a   *b*

8. *a*: Space-filling representations of molecules show how they occupy space. This is the enzyme molecule DNA polymerase, which constructs new DNA molecules. The shading distinguishes different kinds of atom. *b*: Incomplete knowledge of atomic-scale structure, or simply desire to avoid too much detail, can sometimes necessitate very schematic depictions. Here I show the ribosome complex, which constructs new proteins.

There are various ways of accommodating the third dimension, which have become more sophisticated with the advent of computer graphics. Fig. 7 (see page 22) shows a stereoscopic representation of a medium-sized molecule: with practice you can get the two images to overlap and see the shape in 3D.

This by no means exhausts the schemes that chemists have devised. Sometimes there is a call for 'space-filling models' that reveal how great a volume of space the molecule occupies (Fig. 8*a*). Sometimes *ad hoc* schematizations work best, to avoid showing unnecessary detail (Fig. 8*b*).

Making molecules

The molecules in Fig. 8 are biomolecules: fantastic constructions painstakingly assembled by the cell's machinery so that every atom sits in the right place. Chemists cannot yet approach this level of artistry, which is why nature often gets the upper hand. The molecules that we make for killing pathogenic bacteria, for foiling viruses, or for destroying cancer cells often do their job rather crudely. They work, often surprisingly well, but they might damage healthy cells in the process. Or the invading organism might quickly find a way to outwit them, as bacteria are finding ways to develop immunity to antibiotics. Chemists are getting rapidly better at the craft of molecule building, however, and it is not too much to hope that one day drug treatments will be free of side effects, and guaranteed to be successful.

Primo Levi put it this way:

> when you come right down to it, we're bad riggers. We really are like elephants who have been given a closed box containing all the pieces of a watch: we are very strong and patient and we shake the box in every direction and with all our strength. Maybe we even warm it up, because heating is another form of shaking. Well, sometimes, if the watch isn't too complicated, if we keep on shaking, we succeed in getting it together . . .

This is actually something of a worst-case scenario. Levi wrote it in 1978, and things have come a long way since then. Yet, until the last few decades of the twentieth century, the approach that Levi describes, which chemists like to call 'shake and bake', was often the best they could do. Most of the effort in synthetic chemistry is devoted to making so-called organic molecules, which means that they have skeletal frameworks built largely from carbon atoms. Most of the molecules I have depicted so far are considered to be organic molecules. In dimethyl ether and ethanol, the carbon framework is rather small. In Primo Levi's molecule (page 15), it is a more complicated skeleton. And you can see that nitrogen atoms are also part of his framework, while an oxygen atom serves as a crucial bridge in dimethyl ether. Organic molecules are not necessarily based on skeletons exclusively of carbon – just predominantly so.

'Organic' might seem a strange choice of word, for nearly all of the molecules that organic chemists play with are products not of nature's organisms but of the laboratory. The term is a historical one, for organic chemistry was once indeed the study of the molecules derived from living organisms. These, it became clear, were largely carbon-based. Why carbon? Atoms of carbon are almost unique amongst the elements in their ability to link together into stable frameworks with complicated shapes: rings, long chains, branching networks.

Chemists of the nineteenth century had at best only a dim awareness of how to make new organic molecules. They could modify the molecules that nature provided, chipping fragments off the carbon backbone and replacing them with others; but altering the framework itself was more difficult. The problem was rendered still more refractory by the fact that they usually had little idea of the true architecture of the molecule they wanted to make. It is a wonder that their shake'n'bake methods got them as far as they did, providing the first synthetic plastics, dyes, and drugs.

By starting where some of these chemists started, we can begin to see what molecule building is about, and why it is so difficult – and so desirable. In the mid-1850s, the German chemist August Wilhelm Hofmann, working in London, directed his teenaged student William Perkin to make the compound quinine from the distilled components of coal tar. Quinine is a natural extract of the cinchona tree, and was used to treat malaria. Coal tar was a sticky black residue produced in great quantities by gas works, which sprang up in the early part of the century following the invention of gas lighting. It was unpromising stuff, but Hofmann and others had discovered that by distillation one could separate from it several carbon-rich, odorous ('aromatic') organic compounds such as benzene, toluene, xylene, and phenol.

No one knew the structures of any of these compounds – no one could draw stick diagrams like those shown earlier, indicating the connectivity of the atoms. All they knew was how much of each element the compounds contained, and thus what its chemical formula was. Benzene, for example, has the chemical formula C_6H_6, and quinine $C_{20}H_{24}N_2O_2$. The shape of the carbon backbones in these molecules was totally unknown.

Hofmann's procedure (and therefore Perkin's) was to count atoms. They began with a coal-tar extract that they could convert to a compound called allyltoluidine with most of the right atoms in roughly the right ratios, and hoped that some appropriate treatment of this substance would convert it to quinine. They guessed that two molecules of allyltoluidine (with formula $C_{10}H_{13}N$) might combine with some oxygen and hydrogen to make the drug. But it was a long shot, for there are many ways of linking ten carbon atoms together; and, as it happened, the carbon framework of allyltoluidine is not the same as that of half a quinine molecule.

So the experiment, which Perkin conducted in a laboratory rigged up at his parents' house in east London, did not work – it just gave a rust-coloured sludge, something unhappily familiar to organic chemists. Yet the young Perkin did not give up – he tried instead starting with an organic compound called aniline in place of allyltoluidine. This time the sludge was black. But when it was dissolved in methylated spirits, it produced a glorious purple colour, which, to Perkin's delight, would dye silk. He had discovered the first aniline dye. He set up a factory with his brother and father to make the stuff, and it was soon being manufactured in large quantities in Britain and France. This marked the beginning not only of the synthetic dye industry but of the entire modern chemicals industry – for many of today's chemicals companies, such as BASF, Ciba-Geigy, and Hoescht, began as manufacturers of aniline dyes.

By the last quarter of the century, the synthesis of organic molecules had become a less haphazard business. August Friedrich von Kekulé deduced in 1857 that carbon atoms are four-valent – they like to form four bonds. And in 1865 he proposed that benzene, to which all of the aromatic coal-tar molecules were related, contains a *ring* of six carbon atoms – an ubiquitous leitmotif of organic chemistry. In 1868 the German chemists Carl Graebe and Carl Liebermann synthesized the alizarin molecule, which is responsible for the red colour of the dye extracted from the root of the madder plant. This was one of the most commercially important of all natural dyes, and Graebe and Liebermann's synthesis eventually made it available much more cheaply as an artificial product.

The synthesis of alizarin stands as a landmark in molecule making for two reasons. First, it was achieved by planned modification of the starting material (another coal-tar aromatic compound called anthracene), rather than by cooking up the ingredients and hoping for the best. The chemists had some knowledge not only of the formula but also of the chemical structure of anthracene, which they knew to be related to that of alizarin. (In fact they guessed the wrong structure, but fortunately this turned out not to matter.) Organic chemists call this kind of procedure – in which a starting molecule is converted systematically, bit by bit, to the desired product – a *rational synthesis*.

Secondly, by making alizarin in the laboratory Graebe and Liebermann had shown that organic chemistry now had a prowess to rival nature's. It was possible to make the complicated molecules found in living organisms, which chemists today call *natural products*.

So was the synthetic red dye made by Graebe and Liebermann, and later by the ton in chemicals factories, identical to the natural red madder dye? Yes and no. The extract traditionally obtained from madder root is a mixture of several different compounds. Alizarin is the main colourant molecule, but the extract also contains a closely related compound called purpurin, which (despite the name) imparts an orange colour. The process that converted anthracene to synthetic alizarin also generated several side products, mostly molecules with structures very similar to alizarin. Perkin was one of the chemists who, in the early 1870s, identified at least four side products in the alizarin manufactured industrially. There were doubtless many others present in even smaller quantities.

So, whereas the alizarin molecules made synthetically were identical to those extracted from madder root, the actual synthetic dyestuff was different from the natural dye – and both were impure. This is true of just about any 'natural-product' compound manufactured industrially, because all the synthetic procedures used by organic chemists generate side products. It does not necessarily mean that synthetic chemicals are better or worse than the equivalents extracted from natural sources: both are likely to be impure to some degree. But chemists place great value on purity, and spend much time ridding their products of impurities. Natural extracts, in contrast, are complex mixtures unless processed to separate the components.

The commercial value of substances derived from coal-tar compounds was not restricted to dyes. Paul Ehrlich, a German medical scientist, used the new synthetic dyes in the 1870s for staining cells, making them easier to study under the microscope. He noticed that some dyes would kill the bacterial cells that absorbed them, which suggested therapeutic possibilities. He began to synthesize dye compounds to test as drugs, and in this way he found in 1909 an arsenic-containing dye that would kill the parasite that causes syphilis. As Salvarsan, this drug offered the first relief from the deadly disease since the medieval use of mercury. It was the beginning of modern chemotherapy.

Nineteen years later Alexander Fleming discovered penicillin, a compound produced by a mould, which killed bacteria. This was the first antibiotic, and it revolutionized surgical medicine by greatly reducing the risk of wound infection. Many other natural products have beneficial physiological effects: salicylic acid, for example, an extract of willow bark, has both antiseptic and analgesic (pain-killing) properties, and a closely related molecule provides the drug aspirin, manufactured by Bayer since 1899. Chemists and medical scientists continue to comb through nature's arsenal of molecules for potential drugs, and then to find ways of synthesizing those that work.

One of these that has gained fame in recent years is the compound called paclitaxel, better known by the trade name of taxol. It is a natural product of the Pacific yew tree, and was found in the 1980s to be highly effective at preventing cells from dividing. This makes it a potential anti-cancer agent, since cancer is the result of uncontrolled proliferation of cells. It has been approved by the US Food and Drug Administration for the treatment of breast, lung, ovarian, and prostate cancer. But the Pacific yew does not provide a reliable source, since each tree yields up only a few milligrams of the compound, and it has to be extracted from the bark, so that the tree dies. Already an endangered species, it would be wiped out before meeting the global demand for taxol. There is clearly a call for synthetic taxol.

The molecular structure of taxol is fiendishly complicated. Its backbone consists of four rings of carbon: one with four atoms, two with six, and one with eight (Fig. 9). Various subsidiary groups of atoms dangle from this framework. There is no standard chemical reagent available that has this skeletal form – it must be constructed from scratch.

This is a profound challenge for the chemical rigger. Primo Levi's character gives an indication of the way that a synthetic organic chemist today would go about it:

> as you can imagine, it's more reasonable to proceed a bit at a time, first attaching two pieces, then adding a third, and so on. It takes more patience [than shake'n'bake], but actually you do get there first. And most of the time that's the way we do it.

A synthesis like this has to be carefully planned. The most common

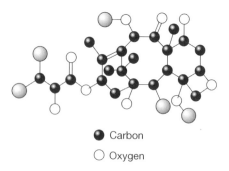

● Carbon

○ Oxygen

9. The taxol molecule. Here I show only the key elements of the framework: the black spheres denote carbon atoms, and the white spheres oxygen. The larger grey spheres are 'substituent' groups containing carbon, oxygen and hydrogen, which I have declined to show in detail. There are other hydrogen atoms in the molecule too, which I have omitted for the sake of clarity.

method of planning is that devised by the American Nobel laureate Elias J. Corey, who called it 'retrosynthetic analysis'. As the name implies, you work conceptually backwards from the finished product, as if disassembling a model of the molecule. At each step you break bonds that you can see how to forge, so that, when it comes to conducting the forward process, you have already seen how each link is to be made. The trick is to work back to starting materials – fragments of the carbon framework – that are readily available, or that can easily be synthesized from off-the-shelf compounds.

In the case of taxol, two groups 'got there first'. A team at the Scripps Research Institute in California, led by K. C. Nicolaou, and a group working under Robert Holton at Florida State University, described multi-step synthetic procedures within a week of one another in 1994. There is no unique way to make a molecule this complex, nor indeed a 'best' way – several alternative pathways have since been reported. But they all remain too complex to be viable for large-scale production, and taxol is currently manufactured 'semi-synthetically': from an intermediate compound found in the

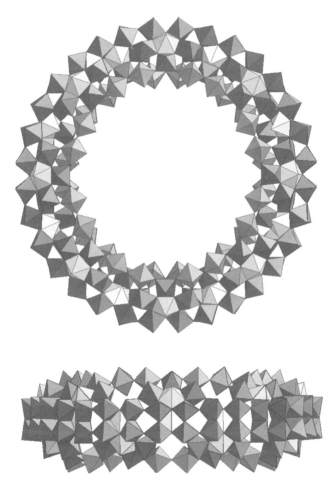

10. A molecular 'big wheel' made by Achim Müller's group in Germany (top and side views). Each pyramid is a cluster of molybdenum and oxygen atoms.

yew needles, which is like half-built taxol. The synthesis can be completed relatively efficiently in the laboratory, and the needles can be removed without killing the trees.

I have talked here about making organic molecules, but I should emphasize that many chemists build molecules based on elements other than carbon. These are often rather small molecules, since other elements do not so readily form the large, intricate frameworks that carbon will adopt. One of the more remarkable exceptions to this rule is the molecule shown in Fig. 10, a ring of mostly atoms of molybdenum and oxygen. It was made by Achim Müller's group at the University of Bielefeld in Germany, and measures fully four nanometres across (that is, about fifteen times the width of a water molecule, and tens of thousand times smaller than the width of a human hair). When metals and oxygen combine, they do not usually stop in the middle ground of large molecules: either they will form molecules of a few atoms each or they will crystallize into mineral-like solids ('infinite molecules', if you like). Chemists have recently become very interested in large *inorganic* molecules like the ones shown here, because they can show unusual and possibly useful behaviour such as magnetism or electrical conduction. Components such as transistors on microchips are made of inorganic materials, primarily silicon and silicon dioxide. Tailor-made components for a molecular-scale electronics technology are just one of the items on the menu of today's sophisticated molecular cookery.

Chapter 2
Vital signs: the molecules of life

It is a little comforting when scientists and poets reach the same conclusions. Pondering the question 'What is life?' in 1949, the British biologist J. B. S. Haldane began by confessing:

> I am not going to answer this question. In fact, I doubt if it will ever be possible to give a full answer, because we know what it feels like to be alive, just as we know what redness, or pain, or effort are. So we cannot describe them in terms of anything else.

Emily Dickinson was more concise:

> Nature is what we know
> Yet have no art to say.

Yet Haldane was prepared to venture further:

> life is a pattern of chemical processes. This pattern has special properties. It begets a similar pattern, as a flame does, but it regulates itself as a flame does not . . . So when we have said that life is a pattern of chemical processes, we have said something true and important . . . But to suppose that one can describe life fully on these lines is to attempt to reduce it to mechanism, which I believe to be impossible.

Is life mere molecules, acting together with awesome but in principle explicable complexity? Or is something more involved? We simply do not know. Scientists take a bottom-up approach, assuming the minimum and invoking only testable hypotheses. Whether this will eventually lead to a point beyond which science is powerless to proceed, we cannot yet say. But no such point has so far become apparent. It seems possible that life – which we might loosely define as an organism that can reproduce, and respond to and extract sustenance from its environment – may be nothing but molecules and their relationships. Indeed, this seems extremely likely. It need not be disappointing; quite the contrary, it would be remarkable. That a conspiracy of molecules might have created *King Lear* is a possibility that makes the world seem an enchanted place.

I do not think it likely, however, that the human mind (let alone the wonders it concocts) will ever be *explained* in molecular terms, any more than *Lear* is explained by the alphabet. Most scientists do not believe so either. Phenomena are hierarchical: all things cannot be understood by considering only what transpires on a single rung. No matter how well I understand the way a transistor works, I will not be able to deduce from this knowledge why my computer crashes. If I sow seeds that fail to grow, I will do better to begin by thinking about the nutrient content, humidity, and temperature of my soil than by performing a genetic analysis of the seeds. Much of the skill in doing science resides in knowing where in the hierarchy you are looking – and, as a consequence, what is relevant and what is not.

It is worth spelling this out before we explore the molecules of the living world, because a molecular view of biology is often branded as reductionistic – as an attempt to explain every aspect of life at the molecular level of genes. This is indeed sometimes the best way to proceed, for molecules are after all the smallest functional units on which life is founded. But if we accept, as most scientists do, that by descending the ladder to the microworld we must forgo a whole range of questions and answers about life (such as: what is consciousness?), then there seems to be nothing obviously objectionable about the descent.

Indeed, this path has led us to a clearer understanding of our fundamental nature. Molecular biology has helped to fill the major gap in Charles Darwin's evolutionary theory: the issue of the *mechanism* of natural selection. It has given us at least some inkling of how life came into being on a planet of gas, rock, and water. It has saved lives and relieved much pain and suffering. It has helped us to understand why medicines do not always work as we might hope, why irresponsible use of antibiotics has bred superbugs, how the AIDS virus does its terrible work. The study of life's molecules became the major science of the twentieth century's second half, and looks set to have an ever greater impact on our lives in the future. It is perhaps the one area of science in which some degree of knowledge is no longer a luxury.

The vital force

Organic chemistry was once considered different from the rest of chemistry. Many scientists believed in the early nineteenth century that organic matter was the product of a vital force operating in living organisms, which the chemist could never mimic in the laboratory. But by 1818 the influential Swedish chemist Jons Jacob Berzelius saw tautology looming in the idea of vitalism, while despairing of ever getting beyond it:

> the cause of most phenomena within the Animal Body lies so deeply hidden from our view, that it certainly will never be found. We call this hidden cause *vital power*; and like many others, who before us have in vain directed their deluded attention to this point, we make of us a *word* to which we can affix no idea.

Yet in the same breath, as it were, Berzelius hinted at how to go further:

> This *power to live* belongs not to the constituent parts of our bodies, nor does it belong in them as an instrument, neither is it a simple power; but the result of the mutual operation of the instruments and rudiments on one another . . .

Here is the key. Understanding the molecular basis of life is not so much about appreciating what the molecules are, as what they do to one another. The molecular nature of life is not a gallery but a dance. In later chapters I shall describe some of the steps; here I want briefly to introduce some of characters.

Molecules

In Haldane's time it was not unusual to regard life as a series of chemical transformations being conducted as if in some vast network of laboratory glassware. The key to it all, scientists believed, was metabolism: how we obtain energy from food. But you will not make an organism by throwing into a pot all of the purified molecular components of the cell. A modern view of molecular biology is concerned with *organization* in time and space. How do the molecules of life arrange themselves amongst the cell's compartments, how are they shifted around, how do they communicate so as to synchronize their action? We can ask these questions only because we can now inspect the working cell at the molecular level, taking measurements and snapshots of molecules going about their business. And so the cell becomes a community.

Yet it is a community of Byzantine complexity. Molecular biology is not difficult in the way that theoretical physics is difficult – the concepts are not unfamiliar, abstract, or mathematically abstruse. The difficulty arises because there is so much going on all at once. We react with surprise and shock when things go wrong with our own molecular machinery, but it is far more astonishing that the machinery works at all. Frequently it does so because it is designed to be robust in the face of the world's vicissitudes. There are checkpoints, safety mechanisms, back-up plans, careful record-keeping. No human-made mechanism is anything like as sophisticated or as well organized as a cell.

It is important to bear in mind that the cell is a community of automata. Its members have no volition, no foresight, no memory, no altruism (nor selfishness, in the strict sense). They often collaborate so beautifully that it is easy to forget this. On the other hand, cells can be unpredictable, because we know so little about how they work. They might survive when we expect them to die, or they might react to a potential drug in totally unforeseen ways.

Molecular biology works at the level of the cell, and seldom talks about the whole organism. The cell is the 'atom of life' – you cannot get any smaller and still be alive. (Viruses are a debatable exception – they are little more than genes wearing a coat, but they cannot reproduce without hijacking the machinery of the cells they infect.) This need not be as restrictive a viewpoint as it might seem, since we can understand an awful lot of our requirements as humans according to what goes on in a single cell. A human cell needs oxygen and sugar to make new molecules and to replicate itself – so we breathe and eat. Nerve impulses start at the level of the cell. Our tissues – skin, hair, bone, muscle – are put together molecule by molecule in the cell. We excrete to remove the cell's waste. We shiver and sweat to stabilize our cells' temperature. Very many questions about the way we function can, in other words, be addressed on the rung of molecular biology. Of course, amongst those that cannot are many of the most interesting.

The players

Haldane attributes to Engels the proposition that life is 'the mode of existence of proteins'. (He was a socialist, and did not generally read Engels for the biology.) This vitalistic statement implies that proteins are inherently alive, an idea that Haldane squashes. But he has no objection to the idea that *proteins* are the stuff of life.

Proteins are substances found ubiquitously in living cells. Many of them are enzymes, molecules that catalyse processes of chemical change. Enzymes speed up chemical reactions by factors of perhaps several millionfold and thereby ensure that the body's chemistry is not impossibly slow. They were discovered from studies of fermentation: *enzyme* is Greek for 'in yeast'. In the late nineteenth century it was found that enzymes could be extracted from yeast cells and purified, yet would remain capable of bringing about fermentation despite the fact that they were no longer part of a living system. This discovery helped to establish that the chemistry of life works according to the same principles as the chemistry of non-living matter.

If the cell is a city, enzymes are the workers. To keep the city running, raw materials are imported and converted into useful items. Enzymes populate the cellular factories in which this is done. One curious aspect of this manufacturing industry is that it includes factories for making the workers themselves: enzymes too are put together on a production line.

Not all proteins are enzymes. Some serve a structural role, providing the tissues of the body. Some act as the cell's police force, others carry packages to and fro in a protein shuttle service running on protein tracks. Some operate the portals to the cell, sitting in the outer membrane and opening or closing in obedient response to the instructions they receive. There are something like 60,000 different varieties of protein molecule in human cells, each conducting a highly specialized task.

It would generally be impossible to guess what this task is merely by looking at a protein. They are undistinguished in appearance, mostly globular in shape (see Fig. 8, page 23) and composed primarily of carbon, hydrogen, nitrogen, oxygen, and a little sulphur. All the proteins that fulfil a particular task have the same shape and structure – the seemingly amorphous blob is in fact exquisitely designed and assembled.

Many enzymes are shaped a little like a knobbly kidney bean, with a cleft in the inner curve. This cleft is where the action is – where the molecule performs its catalysis. Some proteins do their jobs in groups: they become 'subunits' of a many-protein assembly. The enzyme tryptophan synthase, which bacteria possess, is one of these, built from four detachable subunits. This enzyme synthesizes the small molecule tryptophan, which is essential to all organisms. Humans do not possess the enzyme, and so we have to get our tryptophan ready-made by eating organisms that have constructed it.

As this example implies, enzymes and other proteins are commonly given names that reveal their function. Alcohol dehydrogenase is an enzyme that takes a hydrogen atom from ('dehydrogenates') an alcohol molecule. ATP synthase synthesizes the molecule ATP. But not all protein names are so transparent. Haemoglobin, which carries oxygen in the bloodstream, gets its name from the Greek for blood (*haeme*) along with the fact that it is globular. Myoglobin, to which haemoglobin donates its oxygen cargo in muscle tissue, derives the first part of its name from the Greek word for muscle. Other names are more whimsical. Elastin is an elastic protein found in many flexible body tissues, such as blood vessels and vocal cords. Ubiquitin is a protein found just about everywhere in the body, because it plays a central role in the universally necessary process of destroying obsolete proteins.

You would not guess, from looking at pictures like Fig. 8, that a protein molecule is in fact a single chain of small molecules linked together. The chain is folded and coiled on itself so densely that it looks like just a mass of atoms. But close inspection of the structure obtained from X-ray crystallography (see page 18) allows us to follow the strand as it twists and turns through the compact globule. Protein chemists sometimes show this strand explicitly in a different kind of representation of a protein (Fig. 11). Here you can see that the structure is built up from certain repeating features or 'motifs', such as coils, which are called alpha helices, and so-called

11. In this representation of a protein molecule, its folded-chain structure is made explicit. The flat, roughly parallel arrows denote beta-sheet motifs, while the coiled segments bearing chevrons are alpha helices. The molecule is the enzyme phosphotyrosine protein phosphatase from a cow.

beta sheets, where several parts of the strand lie parallel to one another.

A protein's structure can be conceptually decomposed even further. The chain is made up of small characteristic clusters of atoms, joined in a sequence like beads on a string. These clusters were once separate molecules, called *amino acids*. There are twenty varieties of amino acids in natural proteins. In the chain, one amino acid is linked to the next via a covalent bond called a *peptide bond*. Both molecules shed a few extraneous atoms to make this linkage, and the remainder – another link in the chain – is called a *residue*. The chain itself is termed a *polypeptide*.

Any string of amino acid residues is a polypeptide. We can make them ourselves simply by heating up a mixture of amino acids. But we will not make a protein this way. In a protein the order of amino acids along the chain – the *sequence* – is not arbitrary. It is selected (that is, naturally selected, in Darwin's sense) to ensure that the chain will collapse and curl up in water into the precisely determined globular form of the protein, with all parts of the chain in the right place. This shape can be destroyed by warming the protein, a process called denaturation. But many proteins will fold up again spontaneously into the same globular structure when cooled. In other words, the chain has a kind of memory of its folded shape.

The details of this *folding process* are still not fully understood – it is, in fact, one of the central unsolved puzzles of molecular biology. We do know, however, what holds the polypeptide chain in its compact form in a protein molecule. Many parts of the chain are able to form weak bonds with each other, called *hydrogen bonds*. These glue the chain into alpha helices and beta sheets, for example. And some parts of the chain are held together by stronger bonds forged between sulphur atoms, which dangle from the residues of the amino acid cysteine. Some residues are relatively insoluble in water, and these will tend to gather together in the core of the protein globule, surrounded by more water-soluble parts of the chain. So the folded structure that results depends on the character of the various residues, and where they are located along the chain – in other words, on the sequence. You could say that protein molecules are manufactured with their own folding instructions.

How does the cell's machinery 'know', when making a protein, in which order the amino acids should be strung together? This is where DNA (deoxyribonucleic acid) enters the picture. The sequence of every protein in the body is encoded in the DNA molecules that reside in every cell. Proteins do all the work (or most of it); DNA sits passively waiting to be read, when the need for a protein arises.

DNA information is written in a different language from protein information, but the cell can translate between the two. DNA is another string of small molecular units – another *polymer*. But its building blocks are different. Rather than amino acids, they are molecules called *nucleotides*. DNA is a library of protein structures, written in the characters of the *nucleotide sequence* (see page 135). The information for making each protein is, roughly speaking, encoded in a sketch of DNA called a *gene*.

There is clearly some impressive coordination involved here. Whenever a job needs doing by enzymes, the message must get through to the region where the DNA resides. In human cells and those of all other organisms other than the most 'primitive' single-celled bacteria, this is a central compartment called the *nucleus*, which is fenced off within its own membrane (Fig. 12). Organisms with cell nuclei are called *eukaryotes*.

In human cells DNA is packaged up in bundles called *chromosomes*. To make a protein, the stretch of DNA containing the corresponding gene is unravelled and read. In fact, proteins are made not in the nucleus but in a different compartment called the endoplasmic reticulum, a labyrinthine network of membrane channels. The gene is *transcribed* first into a molecule related to DNA, called RNA (ribonucleic acid). The RNA molecules travel from the nucleus to the endoplasmic reticulum, where they are *translated* to proteins. The proteins are then shipped off to where they are needed. So the cell's molecules must be capable of communication and transport.

The whole affair is regulated so that proteins are not being made willy-nilly, but only on demand. If the cell was constantly making all the proteins at its disposal, it would rapidly seize up. A key insight

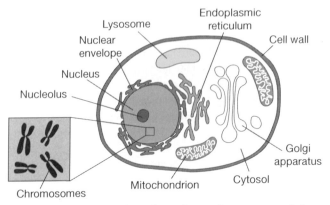

12. Cells of humans and other eukaryotic organisms sequester their genetic material (DNA) in a central nucleus. Various other compartments (organelles) perform a variety of other functions, such as protein synthesis and energy production. Note that the DNA is bundled into chromosomes (as shown here) only when the cell is about to divide; at other times it is unravelled into thin strands.

into how the cell maintains order amongst its component parts emerged from the work of French biochemists François Jacob and Jacques Monod in the 1960s. They showed that genes regulate one another, switching each other on and off via the agency of the proteins they encode. For example, some genes that encode proteins used in the cell (called *structural genes*) are partnered by *regulatory genes* that encode repressor proteins. When the regulatory gene is switched on, the repressor protein is synthesized and binds to the structural gene, preventing it from being 'expressed' (transcribed and translated into its protein). Jacob and Monod called these regulated stretches of DNA *operons*. They provide just one illustration that the cell employs a weblike network of interactions between different genes and proteins. Molecular biologists have now decoded more or less all of the nucleotide sequence of human DNA – but they have so far mapped out only very small regions of the web that it weaves.

Genetic evolution

The molecular science of genes has opened up a Pandora's box. Not only has it helped to explain life's innermost riddles, but it has posed challenging questions about human behaviour and ethics, and offered controversial new technologies. It has also revolutionized our understanding of evolution, and of how we came to be here.

Genes are the currency of inheritance: they are an inevitable bequest from our parents. The idea that characteristics are passed on from parent to offspring is a very old and obvious one, but it was made more concrete by the work of the Austrian priest and biologist Gregor Johann Mendel in the nineteenth century. His experiments on the propagation of peas led him to suppose that there are 'particulate factors' that mediate heredity, passing from the cells of the progenitors to their progeny. It soon became clear that these 'factors', later called genes, were molecular in nature, but throughout the first half of the twentieth century many scientists thought they were protein molecules. Haldane, as we have seen, shared this belief. Not until Francis Crick and James Watson deduced the structure of DNA in 1953 was there a compelling case for regarding DNA, rather than proteins, as the molecular stuff of heredity, the fabric of genes.

This placed evolution on a concrete molecular basis – for what a fertilized egg gets from its parents is not a preformed body but a set of genetic instructions for a body plan. The evolutionary changes that happen slowly from generation to generation are due to changes in the molecular make-up of the genes. DNA is copied when a cell divides – but not always perfectly. So the DNA that a child gets from mother and father may be a slightly flawed amalgam of both their genes. Generally these flaws will not matter. Sometimes they will be harmful (but note that most genetically based diseases are the result of a child *inheriting* a faulty gene, not acquiring one from random copying errors). Very rarely, a genetic mutation will have a beneficial effect, making the organism better equipped for survival. The advantage might be extremely slight – but evolution advances through such infinitesimal steps, as tiny advantages lead to fractionally higher reproductive success and thus to a slow increase in the incidence of the mutated gene in the population.

What this all means is that genes are a molecular record of evolution. The common ancestor of humans and rabbits shared the same set of genes. The differences that now exist between the respective total complement of genes (the *genomes*) of humans and rabbits reflect the divergence attributable to an accumulation of genetic mutations. This enables scientists to reconstruct evolutionary histories – to deduce the order in which species diverged – from the molecular structure of genes. Previously, palaeontologists had to make such deductions on the basis of the shapes of bodies or bones; now they have a more readily quantifiable molecular measure of evolutionary change.

In particular, evolutionary trees or *phylogenies* can be reconstructed by comparing the special DNA that is housed in a compartment of cells called the mitochondrion (see Fig. 12). This is the cell's furnace, where the energy is produced (see Chapter 4). Unlike the DNA in the nucleus, mitochondrial DNA comes directly from the mother. Whereas nuclear DNA is changed by reshuffling the genes of both parents, mitochondrial DNA is changed only by the gradual accumulation of mutations from generation to generation – so it is a better record of evolutionary change. In 1987 Allan Wilson of the University of California at Berkeley and co-workers compared the mitochondrial DNA of people from many racial groups. Knowing the average mutation rate, they figured out that all of the samples, and by extension the mitochondrial DNA in all living humans, derived from a single version that existed 200,000 years ago in the cells of an African woman – the common ancestor of all of humanity. Molecules contain a record of history richer than anything to be found in fragments of clay pots or ancient burial mounds.

The RNA world

All living organisms, from the humblest of bacteria to the most regal of kings and queens, have their genetic material packaged in DNA, and put into effect by proteins. This implies that all life has a common origin.* The most primitive single-celled organisms must have contained proteins and DNA very similar to those in 'simple' bacteria today.

But what came before? The molecular symbiosis whereby DNA encodes proteins and proteins help DNA to function and to replicate is fantastically sophisticated even in a bacterium. Neither proteins nor DNA could have come spontaneously into existence from fragments of organic molecules scattered throughout the seas and lagoons of the early Earth: their structures are just too complex to have assembled at random. It is easier to understand (at least in principle) the 3.8 billion years of evolution from the earliest bacteria or algae to the present day than to understand how, over maybe just a few hundred thousand years, Earth changed from a barren planet to one that cradled life.

Chemists have devised many inventive schemes according to which the inorganic constituents of the young Earth, such as methane, carbon dioxide, ammonia, water, and nitrogen, might have become transformed into the amino acids and sugars needed to make the molecules of life. They are all tentative; no theory yet prevails for the chemical origin of life. But the conceptual difficulties in progressing from rock, gas, and water to prototypes of biomolecules

* We simply do not know whether the DNA–protein partnership is the *sine qua non* of life, and it would be rash to suggest that life can have no other molecular basis. But none has been found. It is not so difficult to imagine that slight, systematic modifications to DNA could give rise to an alternative genetic system – but no organisms show such a thing.

are still smaller than those of turning these building blocks into functioning cells full of proteins and DNA. It is a chicken-and-egg problem: on their own, both proteins and DNA are useless.

The favourite way out of the conundrum is to switch attention to the humble go-between: RNA, which carries the genetic information to the machinery of protein synthesis. RNA is much more versatile than DNA. In the 1980s, RNA molecules in the cell were discovered that could act as catalysts for their own rearrangement. Human genes are corrupted with a lot of 'junk' that needs to be excised before the message can be clearly read (see page 140). This junk gets copied into RNA, but is then snipped out before the RNA is translated into a protein. This editing is largely conducted by enzymes; but some RNA molecules can do it unassisted. These are called *ribozymes*, reflecting the fact that they show enzyme-like tendencies.

During the 1990s, biochemists greatly expanded their appreciation of RNA's abilities. Using the techniques of biotechnology developed for manipulating and rewriting DNA, they have made synthetic RNA molecules that can conduct all manner of chemical processes, such as linking together nucleotides or forming bonds between carbon atoms. These studies show that in principle RNA is versatile enough to bring about many of the chemical transformations that would have been necessary for life to begin. In short, RNA can act both as a gene-carrier and as a worker.

Many scientists researching into the origin of life therefore postulate an era that they call the *RNA world*, which existed before the double act of proteins and DNA appeared. RNA is still a fiendishly difficult molecule to make under conditions comparable to those on the early Earth. But the RNA world breaks the impasse posed by the mutual dependence of proteins and DNA, and so provides a conceptual link between the formation of small organic molecules and the appearance of the first primitive cells.

Synthetic life

If we do eventually come to understand the origin of life – not necessarily how it really began, but at least how it *might* have begun – could we then rerun it in the laboratory? Can we create life from scratch?

Enough is known already about the molecular basis of life for researchers to be able to speculate about building an artificial cell. It sounds perhaps like a frightening prospect. What if we happened to make a cell that was far better at replicating than 'natural' cells? Would they colonize the planet like an alien invasion?

This is not science fiction. In fact, I think that a synthetic cell will be made within the twenty-first century, for better or worse. It is already routine to make synthetic DNA and to rewrite genes, and many chemists are working on building 'designer proteins' from scratch. The biochemists Jack Szostak, David Bartel, and Pier Luigi Luisi have proposed that:

> Advances in directed evolution and membrane biophysics make the synthesis of simple living cells, if not yet foreseeable reality, an imaginable goal.

They suggest that a 'minimal cell' could be constructed from tailor-made ribozymes. A primitive version of a ribozyme that can assemble RNA (and so potentially replicate itself) has already been reported. These could be encapsulated within artificial membranes like those of cells, but with an ability to grow and divide: Luisi has made such 'replicating membranes'. The replicating 'protocells' might evolve RNA molecules capable of assembling amino acids into proteins. This would then allow us, say the researchers, to 'replay the tape of early evolution'.

Those who foresee terrible purposes in such experiments might bear in mind that this is an absurdly difficult way to try to develop some deadly weapon, when lethal chemical and biological weapons can be made already with relative ease. All the same, it is impossible to know where such research might lead. That is the reality of molecular science: it is a creative discipline, which gives us something to show for our efforts at the end. Therein lies all the artistry, all the wonder, and all the peril. In the end, we will only get the molecules we deserve.

Chapter 3
Take the strain: materials from molecules

The hardest part of space travel (apart from the boredom and the danger) is the leaving. In the vacuum of space, free from strong gravitational influences, a small burst of propulsion will keep a rocket moving almost indefinitely. So most of the fuel that a rocket takes on board is needed simply to escape the Earth's gravity. This fuel and the engines that burn it account for the greater part of a rocket's mass. I recall vividly the Apollo missions that left the Earth as a gleaming tower and returned as a tiny nub from the nose.

If we could launch spacecraft from outside Earth's atmosphere, the payload would therefore be much diminished. In his novel *The Fountains of Paradise*, Arthur C. Clarke suggested how this might be done. He posited the Space Elevator: a platform positioned in geostationary orbit around the Earth, tethered to the ground by a long, superstrong cable. Space hardware and any passengers are shuttled up by elevator to the platform, from where they can be launched into space with a fraction of the fuel requirements needed for a ground-based takeoff.

To tether an orbiting platform to the Earth's surface, we would need cables far stronger than anything currently available. They would need to be lightweight too – the weight of that much steel cable would be immense.

It does not take the speculative concept of a Space Elevator to explain why we need materials that are strong, tough, corrosion-resistant, lightweight, and so forth. But scenarios like this serve to motivate the question of just how far one can go in improving materials' properties. Superstrong 'space tethers' are also being considered for launching objects into space by a kind of slingshot process in which the payload is temporarily attached by a thread to an orbiting satellite. Such tethers have to be both lightweight and strong. And there is no lack of more mundane demands for strong cables: for example, to hang bridges and tether drilling rigs to the seabed.

Tough fibres have always been available to us, for we are fortunate that nature supplies them in abundance: silk, hemp, wood, hair. In pre-revolutionary Russia silk was used for bulletproof protection, and artificial silk is being developed today with the same purpose in mind.

The plastic age, which began in earnest in the early twentieth century, has supplemented natural fibres with synthetic ones that have both advantages and shortcomings. The earliest plastics were made by trial and error; modern plastics, in contrast, are designed at the molecular level for the applications they will serve. In this chapter on the molecular aspects of materials, I will focus my discussion on fibres both natural and human-made, since these provide some particularly beautiful and graphic examples of how molecular structures affect the kind of material properties that engineers worry about.

But the interaction of molecular science with materials engineering is far, far broader than this. Molecules are designed that can be converted into ultrahard ceramic materials capable of withstanding high temperatures. These are used in aerospace engineering, turbine manufacture, and power generation. Materials made from molecules (particularly polymers) are designed to conduct electricity, to capture, guide, and transform light pulses, to sieve other molecules, to protect surfaces from corrosion or contamination. They include 'smart' materials that respond to changes in their environment, which can act as autonomous switches, valves, and pumps. The impact of molecule-based materials on medicine is huge, and destined to be ever-more important: they provide artificial limbs, organs and tissues, systems for administering drugs, biodegradable sutures for surgery, sensors for monitoring the body's state of health. One day molecular engineering will make it possible to grow a new kidney or a new heart to replace a damaged one.

Cables of the body

The idea that our bodies are a mass of cellular communities of proteins does not tally with our experience. We experience ourselves as a composite of fabrics: skin, bone, muscle, hair, fingernails. This material framework has the properties necessary to let us interact with the world. The outer layer of skin is just cladding, made up of cells that are programmed to die once they have formed the tissue. The same is true of hair, which provides insulation, and of fingernails, the evolutionary remnants of claws that once punctured and ripped. Bone and tooth are composed largely of a hard, inorganic material: calcium phosphate. Most of these materials are continuously renewed while we live; some, such as the proteins in the eye's lens, are not.

These natural materials of the body serve mechanical roles and maintain our structural integrity. They are like the bricks, girders, and cladding of a building, which protect the workers from the elements and house all the complex wiring and plumbing that is necessary to conduct business as usual. Many of the body's structural fabrics are proteins. Unlike enzymes, structural proteins do not have to conduct any delicate chemistry, but must simply be (for instance) tough, or flexible, or waterproof. In principle many other materials besides proteins would suffice; and indeed, plants use cellulose (a sugar-based polymer) to make their tissues. Yet the marvel of proteins is that they are so versatile. The molecular chains can be woven into strong fibres; cross-linked or entangled, they form the stiff matrix of horn and claw, or elastic sheets. What is more, the raw materials for making proteins are abundant in the cell. And because proteins are encoded in genes, the molecular features that give a structural protein its mechanical properties can be delicately tuned and then reliably reproduced.

The most abundant structural protein in the human body, comprising about a quarter of our total protein mass, is collagen. This is a relatively simple protein whose chainlike molecules contain mostly two sorts of amino acids: glycine and proline. Glycine constitutes every third link of the chain, with proline and other amino acids (particularly lysine) in between. Some of the proline units are chemically modified, having an oxygen atom added. This takes place in a reaction that involves vitamin C, which is why this compound is needed to maintain healthy tissues. Lack of vitamin C leads to the condition known as scurvy, caused by damaged collagen that has not been replaced.

Collagen exemplifies the way in which natural protein-based structural materials differ from most synthetic polymer-based plastics. Both are composed of chain molecules; but in structural proteins these chains gather together in complex arrangements, forming thicker fibrils like ropes woven from string woven from thread. This kind of arrangement, in which structural elements are built up at successively larger scales from smaller components, is said to be *hierarchical*. Engineers have learnt to use the same principle: Gustav Eiffel's iconic tower, for instance, contains struts made of a network of smaller girders, some of which are built up from even smaller girders (Fig. 13).

Collagen displays several different kinds of large-scale structure,

13. Structural hierarchy, common in nature's materials, is exemplified in the Eiffel Tower.

like several different designs of tower, but all are constructed from the same basic small-scale elements (Fig. 14). Each collagen molecular chain crimples up into a helix. Three of these twist around one another to form a ropelike triple-helical 'microfibril'. These microfibrils aggregate together in various ways. For example, they can gather in a staggered arrangement to form thick strands called banded fibrils. The staggering creates the appearance of dark bands under the electron microscope. Banded fibrils constitute the connective tissues between cells – they are the cables that hold our flesh together. Bone consists of collagen banded fibrils sprinkled with tiny crystals of the mineral hydroxyapatite, which is basically calcium phosphate. Because of the high protein content of bone, it is flexible and resilient as well as hard.

Collagen fibrils are, however, not outstandingly strong on their own, because the molecules are not linked or entangled together. But other types of collagen contain cross-linked bundles of microfibrils, joined into a kind of tough web or mesh. This provides the

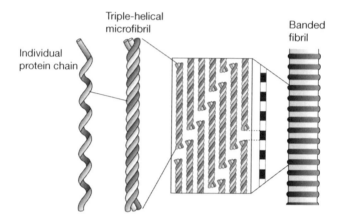

Individual
protein chain

Triple-helical
microfibril

Banded
fibril

14. Collagen has a hierarchical structure of coiled coils. A staggered arrangement of these collagen microfibrils causes the dark bands seen microscopically in banded fibrils, where a metal staining agent attaches itself to the ends of the microfibrils.

membrane separating the outer layers of skin from the inner layers.

In contrast to the disorderly tangle of connective tissue, the eye's cornea contains collagen fibrils packed side by side in an orderly manner. These fibrils are too small to scatter light, and so the material is virtually transparent. The basic design principle – one that recurs often in nature – is that, by tinkering with the chemical composition and, most importantly, the hierarchical arrangement of the same basic molecules, it is possible to extract several different kinds of material properties.

Collagen also constitutes the tough, elastic fabric of tendons and ligaments, and the reinforcing dentine of tooth. But the proteins on our outside – in skin, hair, and nail, as well as animal horn and hoof – are of a different kind. These tissues consist mostly of keratin, another hierarchically structured fibre. The molecular chains of keratin are again wound into helices, which are paired up in double-helical coiled strands. Two of these strands are wound together in a 'supercoil' called a protofibril, and the primary cables of keratin are composed of clusters of eight protofibrils. These fibres are surrounded by a matrix of disorderly keratin-like proteins cross-linked by sulphur atoms, like steel cables embedded in concrete. The cross-links determine the strength of the material: hair and fingernail are more highly cross-linked than skin. Curly or frizzy hair can be straightened by breaking some of these sulphur cross-links to make the hairs more pliable.

Hair is a useful natural fibre; but most of the materials made from collagen and keratin proteins are formed instead into sheets (such as skin) or lumps (such as horn and hoof). They have a fibrous structure at the molecular and the microscopic scales, since it is easier for the production machinery of individual cells to manufacture such structures than, say, to cast solid blocks. But the body then organizes these micro-fibres into other shapes.

Web dreams

Silk, on the other hand, shows just how impressively evolution can rise to the challenge of making fibres when they are truly called for. Here is a substance that can be spun into threads all but invisible to the passing fly, yet flexible enough to absorb the energy when the fly hits the web, and strong enough not to break under the impact (Fig. 15). Stronger than steel or the best human-made fibres, silk is admired by the engineer for its robustness and prized by the textiles manufacturer for its exotic shimmer, its cool texture, and its ability to soak up bright dyes.

The spider makes silk for many uses, and gives it a different flavour in each case. The web is spun from dragline silk; other silks are used to make supporting fibres, threads that attach the web to the branch or the rafter, strands to bind prey, strands to swaddle the developing larvae, and so forth. All of these silks are composed of protein

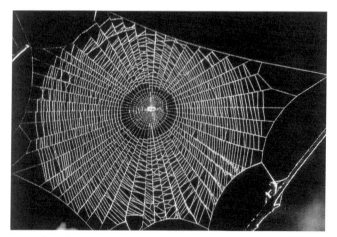

15. Spider silk is one of the strongest known fibrous materials.

chains in which the amino acids glycine, alanine, and serine dominate. But the precise blend of components is tailored to the silk's use.

The wondrous properties of silk are a consequence of the way its protein chains are organized. Whereas in collagen and keratin the chains are coiled into helices, in silk the basic organized structural element is not a coil but a sheet. Neighbouring chains sit side by side in aligned ranks, and each chain is linked to those on either side via relatively weak hydrogen bonds (see page 41), which zip the chains together into beta sheets (Fig. 16).

The orderly, relatively rigid sheets can stack on top of one another to create tiny, three-dimensional protein crystallites. In silk fibres these crystallites are microscopic, extending perhaps only twenty-millionths of a millimetre in any direction. Beyond the crystallite regions, the protein chains continue into less orderly regions, where they are tangled together. So silk is really a kind of composite of tiny crystals dispersed in a more flexible protein matrix – a little like

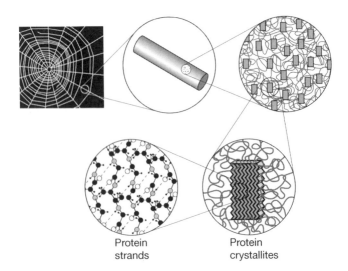

16. **There are several levels to the hierarchical structure of silk. At the molecular scale it is organized into orderly (crystal-like) beta sheets of parallel strands. The dashed lines denote hydrogen bonds.**

Protein strands

Protein crystallites

bone, except that the crystals are now not minerals but proteins themselves.

The crystalline regions of the silk protein generally have a regularly repeating sequence of amino acids. In the cocoon fibres of the silkworm *Bombyx mori*, for instance, the sequence glycine–alanine–glycine–alanine–glycine–serine repeats along the chains. In the disorderly regions, the sequence is irregular.

Silk fibres are insoluble in water – they would be of little use if they dissolved in the morning dew. Yet the spider spins the threads from a solution of protein molecules in water, thereby achieving the seemingly miraculous feat of turning a soluble molecule into an insoluble one. This change in solubility is a result of a change in the way that the chains are organized. The spider manufactures silk protein in its silk gland, at which point it is still soluble. This solution passes from the gland towards an outlet in the body called the spinneret. As it makes this journey, the silk solution loses water and becomes more concentrated, and the chains begin to get zipped together by hydrogen bonds. By the time the silk leaves the spinneret, most of the water has been squeezed out and the chains form beta sheets. Once the molecules are closely packed together in these crystalline regions, it is hard for water molecules to penetrate between them, and so the silk fibres are essentially solid and insoluble.

Molecular scientists cannot yet design artificial polymers with this sort of hierarchical structure, since it is very hard to control the way that molecules pack together over several different size scales at once. The molecular structure of synthetic polymers – which atoms they contain, and in what order and what spatial arrangement – can now be specified quite accurately; but it is another matter to 'program' these molecules to gather into particular kinds of cluster, or to cross-link them at definite locations.

Nevertheless, polymer chemists now know many tricks for engineering particular properties into their materials – making them hard or soft, say, or capable of forming strong fibres. In 1839 Charles Goodyear discovered how to cross-link the polymer molecules of the gum extracted from the tropical brazilwood tree *Hevea brasiliensis* by heating it with sulphur. This so-called vulcanization process converts the soft gum into the stretchy form we call rubber.

Rubber gum is made up mostly of a hydrocarbon polymer called polyisoprene, the chains of which are composed only of interlinked carbon atoms with some hydrogen atoms attached. Many of the plastics now manufactured on a large scale, such as polyethylene, polypropylene, and polystyrene, are also hydrocarbon polymers, made from the products of oil refining. But the rubber of truck tyres is natural, since it remains too hard to make it synthetically.

Until the mid-twentieth century the linking together of small hydrocarbon molecules to form synthetic polymer chains was a haphazard process, producing many different types of chain: some short, some long, some branched, some straight. Because the structure of the chains determines how they pack together, and because this packing determines the properties of the bulk material, this lack of control meant that polymer chemists could do little to fine-tune these properties. Whereas delicate structural control allows nature to make several apparently different fabrics from the same raw material, chemists just got the same kind of plastic every time. In the 1950s, special catalysts were developed that provided greater control over the molecular structure of the chains, and therefore over the way that they pack. This meant, for example, that polyethylene could be prepared in a new, hard, high-density form with a wider range of uses than the older, softer, low-density form – for example, as stiff drums and bottles for packaging, or as pipes and sheets in engineering and construction.

These two forms differ in the degree of orderliness with which the chains are packed together; that is to say, their degree of crystallinity. As silk demonstrates, greater crystallinity results in a tougher, stiffer material, as well as making it denser and less soluble. These are desirable attributes for strong fibres. The high strength arises because molecular chains that are packed in a closer, more orderly manner cohere more avidly.

So one of the principal challenges in making strong polymer fibres is to enhance the alignment of the molecules. Just about any polymer made of straight-chain (that is, non-branched) molecules has the potential to align its chains. But in silk this is helped by their tendency to 'zip' together to form crystalline beta sheets. In other words, the chains themselves have a kind of 'alignment instruction' programmed in. Can we make artificial polymers like this?

Indeed we can. The silk protein chains line up because of the forces of attraction between the units on different chains, which allow the chains to lock together like zips. Similar interactions exist between the chains of a class of synthetic polymers known as aramids, from which the renowned Kevlar fibres produced by DuPont are made.

Kevlar is one of the best candidates so far for tethering a Space Platform. It has a tensile strength greater than that of steel, but is much lighter. The fibres are used as the strengthening cord in rubber tyres, as a fabric for bulletproof clothing, as a reinforcing material in hard composites used for aerospace engineering, and even as woven cables for anchoring oil-drilling rigs.

But gram for gram, silk is stronger still. The high degree of chain alignment in a silk strand does not simply come from the fact that the molecules tend to zip together in solution. As the protein solution flows down a passageway from the spider's silk gland towards the spinneret, the polymer chains line up with the direction of liquid flow, in the same way that our hairs can be pulled into alignment by the wind. The same process operates also during extrusion of the thread from the spinneret. It is called shear-induced alignment (since the flowing liquid experiences a so-called shear force).

Thus, even though it is possible to make silk-like proteins artificially, making silk threads from them is another matter entirely. If a fibre is mechanically extruded from a solution of (natural or artificial) silk protein just like pulling a thread from tacky glue, the fibre is still not as strong as real silk thread. Some researchers believe that only by building a tiny machine that mimics the spider's spinneret will we be able to make artificial silk that compares favourably with the natural material.

Shear-induced alignment is used in an industrial process to make extremely strong fibres from polyethylene. The fibres are drawn from a gel-like substance in a complex procedure that results in a very high degree of alignment of the polymer chains. These fibres, sometimes called 'rocket wire', are even stronger than Kevlar, and, unlike many organic materials, they are chemically very stable. This makes them well suited for use as long-term surgical sutures.

Material genes

One way of making artificial silk polymer is to import silk-making genes into bacteria using biotechnological techniques. Just like any other protein, silk is genetically encoded in DNA: the sequence of its amino acids along the chains is determined by a corresponding sequence of nucleotide units in a gene in the spider's chromosomes. In other words, the spider possesses a molecular blueprint for making the polymer. (But notice how, because of the complexities of the spinning process, this blueprint alone is not enough to make good silk threads!)

The ability of living organisms to define the molecular composition of a polymer with complete accuracy is an enviable one. Modern synthetic techniques allow chemists a great deal of control over the composition of a chain – for example, they can make hybrid polymers by grafting side chains of one chemical type onto a main chain of another, or by alternating between blocks of one type of unit and blocks of another along a single chain. But making a polymer perhaps thousands of units long in which many different units recur in a sequence of arbitrary complexity – and yet with every molecular chain in the material being identical – is something far beyond our current synthetic capabilities. In a linguistic analogy, our state-of-the-art polymers read something like this: aaaaaaabbbbbbbaaaaaaabbbbbbbaaa. . . Nature's polymers, meanwhile, are more like this entire sentence, and pregnant with meaning.

Biotechnology allows the genes from one organism to be snipped out of the respective stretch of DNA and pasted into the DNA of another organism. The recipient then (all being well) treats the new gene as if it were its own, and uses its transcription and translation machinery to make the corresponding protein. One of the potentially valuable and, I think, less controversial prospects of biotechnology is to transfer human genes into bacteria, which can then be bred in fermentation vats to make proteins needed for medicine. But some polymer scientists have realized that this is also a way to make protein-based materials.

As well as transferring real silk genes into bacteria,* one can 'write' artificial genes and express them this way. Structural proteins typically have repetitive amino-acid sequences, since the fibres they form are generally quite uniform along their length. It is relatively easy to make synthetic genes with repetitive sequences, since one need only make the individual repeat units and then join several of them together. (There are ways of ensuring that the resulting synthetic gene has only a specified number of copies of the repeat unit.) Using this idea, some researchers have begun to make genetically engineered protein-based materials in which the chains are programmed to adopt particular structures. 'Designed' silk-like materials have been made in this way, as well as protein-related synthetic materials similar to collagen and elastin, an elastic protein in skin. Hybrid materials can be envisaged that marry a natural protein (such as an enzyme) to a synthetic protein-like chain designed to behave as a material – for example, to pack into insoluble beta-sheet crystallites that will form a water-resistant surface coating. A possible attraction of protein-like materials for medical applications is that they would be biocompatible and biodegradable.

The cell's skeleton

Cells are laced with a network of fibres more rigid than the rope-like structural proteins collagen and keratin. These are called microtubules, and they provide the rod-like tracks on which molecular engines ferry packages around the cell. Microtubules also

* There is also ongoing work to obtain silk protein from goat's milk, by transferring silk genes into goats.

provide a scaffolding on which a cell alters its shape – for example, allowing an amoeba to extend a pseudopod. The hairlike cilia appendages that protrude from cells in our respiratory tract to push dirt-filtering mucus around, and the whiplike flagella that bacteria use to propel themselves with screwlike motions through a fluid – these too are microtubules.

As the name suggests, they are hollow, tubular structures. They are made up of a protein called tubulin, which is not fibrous but compact and globular. Tubulin consists of two nearly identical molecules, which pair up in a kind of dumbbell. The dumbbells stack together like bricks in a cylindrical chimney (Fig. 17).

Molecules

Tubulin

17. Tubular protein filaments called microtubules comprise the scaffolding of cells.

Tubulin units can attach themselves to and detach from the ends of microtubules, by which means the fibres are lengthened or shortened. An amoeba can retract its 'leg' simply by disassembling the microtubules that pushed it out. This ease of assembly enables microtubules to play a central role in cell division. Once the dividing cell has made copies of its chromosomes, it creates two clusters of radiating microtubules called asters. When the microtubules emerging from one aster meet those from the other, they merge at their ends, forming a bundle of bridging fibres between the focal points of the two asters, called the mitotic spindle (cell division is called mitosis). The chromosomes are attached to this spindle and are pulled apart so that half of each is dragged to either pole (Fig. 18). In this way the mitotic spindle provides a structure on which the duplicated genetic material can be separated out into two complete sets. The cell is then stretched and split into two halves on the framework of microtubules, each half containing a full complement of chromosomes.

18. The mitotic spindle is composed of microtubules. It constitutes the framework on which chromosomes (visible here in the centre of the bundle) are arranged and sorted during cell division.

The anti-cancer drug taxol (see page 29) works by disrupting the assembly of the mitotic spindle and so arresting the proliferation of cancer cells. It prevents the disassembly of microtubules, which is essential as they search blindly for one another from the two asters. Unfortunately, taxol has this same effect on healthy cells; but, since cancer cells divide more rapidly, they are the hardest hit.

Carbon pipes

Until the 1990s, little serious thought had gone into the idea of making strong synthetic fibres from tubular molecules. It is easy enough to join small molecules together in chains, but tubes look much more difficult to orchestrate.

But in 1991 a Japanese microscopist named Sumio Iijima, working at the NEC Corporation in Tsukuba, discovered a kind of tubular molecule that could furnish the strongest fibres known. Called carbon nanotubes, these structures literally assemble themselves from a vapour of atomized carbon.

Iijima was investigating a technique used to make carbon molecules called fullerenes, in which dozens of carbon atoms are joined together in hollow cages. The first fullerene, a sixty-atom cage called buckminsterfullerene, was discovered in 1985. Its curious name derives from that of the US architect Richard Buckminster Fuller, who pioneered the 'geodesic dome' made from hexagonal and pentagonal facets. Buckminsterfullerene, or C_{60}, has this same structure at the molecular scale – hexagonal and pentagonal rings of six and five carbon atoms respectively, linked into a spherical cage (Fig. 19).

In 1990 a method was discovered for making fullerenes in large quantities for the first time. It involved passing an electrical discharge between two rods of graphite (which is pure carbon). The energy of the spark vaporized some of the graphite, and the carbon

19. The C_{60} molecule is approximately spherical, and is composed of hexagonal and pentagonal rings of carbon atoms.

atoms joined together, as the vapour cooled, to form C_{60} and other carbon cages. But when Iijima used slightly different conditions in his fullerene generator, he found something new when he examined the sooty debris under the electron microscope.

The soot was full of needle-like objects, just a few nanometres in diameter. On closer inspection, Iijima saw that these were hollow cylinders of carbon, and that each one contained several cylinders nested inside one another like Russian dolls (Fig. 20). These objects became known as carbon nanotubes. No one had previously suspected, even after the discovery of fullerenes, that carbon atoms could arrange themselves spontaneously in this way.

The walls of nanotubes are made of pure carbon with the atoms arranged in sheets of hexagonal rings. This is the same structure as graphite, except that in nanotubes the sheets are curled up into cylinders. At the tube ends, the sheets are kinked into flat-faced end caps. We normally think of graphite as a weak material – after all, it is used in pencils because simply dragging it over paper rubs off

20. The tapering tip of a carbon nanotube, seen in cross-section in the electron microscope.

some of the black carbon. But this weakness is the result of the flat sheets of carbon being only loosely bound to one another, so that they can slide across each other. The individual sheets, in which the atoms are tightly bound, are predicted to be extremely strong – comparable, in fact, to diamond. When these graphite-like sheets are partially joined together by chemical bonds, as they are in conventional carbon fibres, the material gains a great deal in both strength and stiffness.

Carbon nanotubes are potentially the ultimate carbon fibres – the tubes are seamless sheets of graphite-like carbon. They are predicted to have a greater tensile strength than diamond; greater too than Kevlar, silk, or any other fibre natural or artificial you care to mention. Realizing this, Richard Smalley of Rice University in Houston, one of the discoverers of fullerenes, proposed that, if the Space Elevator were to be built, carbon nanotubes would hold it to the Earth.

But there is a snag. So far, carbon nanotubes cannot be grown to a length of more than a fraction of a millimetre. That is not a very long tether. It also makes measurements of the true strength difficult to conduct, although some experiments at the microscopic scale have already tentatively confirmed that these tubes are indeed very strong and stiff. (As strong and stiff as diamond? That is still uncertain.)

To make useful cables from carbon nanotubes, we will need to develop a way of controlling their growth so that the tubes can be extended indefinitely. In the synthesis of normal polymers, there is a technique called 'living polymerization' that allows the chain growth to be temporarily arrested and restarted at will, so that more and more units can be attached to ever-lengthening chains. If someone were to develop a similar process for carbon nanotubes, it would be revolutionary. But so far there is only a hazy understanding of how the tubes grow, making it hard to see how this could be controlled.* For the present, the Space Elevator will have to wait.

* A tentative first step has been reported by French researchers Philippe Poulin and colleagues at the University of Bordeaux I in Pessac. The team made nanotube fibres and ribbons by injecting nanotubes, suspended in water by the action of soaplike molecules, into a viscous liquid polymer. The nanotubes line up as they are injected from a capillary, in much the same way as silk proteins become aligned as they leave the spider's spinneret. The aligned tubes stick together in fibres that can be dried and handled. But these strands are not made from *continuous* nanotubes, and so they are not yet as strong as conventional carbon fibres, let alone diamond.

Chapter 4
The burning issue: molecules and energy

Imagine if the motor car was like the human body, so that its top speed was sustainable only for short sprints. No more driving down the motorway at (let's be honest) 80 miles an hour – that could only be managed for a half-mile trip to the shops. The further you had to go, the slower the speed, so as not to wear the poor vehicle out. (Perhaps you, like me, have driven cars that do seem to behave in this way?)

Superficially, cars and humans are more similar than one might think, and not for the reasons Flann O'Brien would suppose.* Both obtain energy by burning energy-rich fuel in oxygen; and both release an exhaust of carbon dioxide. Yet cars do not get tired when moving at close to top speed for long periods of time. Provided that you keep refilling the tank, they can keep it up almost indefinitely, whereas no sprinter could sustain their speed for an entire marathon even if they were continually chewing on glucose tablets. Why not?

The question leads us into the molecular mechanisms that power the body, which are known as metabolic processes. In many ways, it is metabolism and not replication that provides the best working definition of life. Evolutionary biologists would say

* If you don't know what I mean, read *The Third Policeman* and find out!

that we exist in order to reproduce – but we are not, even the most amorous of us, trying to reproduce all the time. Yet, if we stop metabolizing, even for a minute or two, we are done for.

Feel your hand. It is warm. (If it does not feel that way, try your armpit, or your tongue.) We are typically warmer than our surroundings. Whether waking or asleep, our bodies stay close to a healthy temperature of 37 °C. There is only one way of doing this: our cells are constantly pumping out heat, a by-product of metabolism. Heat is not really the point here – it is simply unavoidable, because all conversion of energy from one form to another squanders some of it this way. Our metabolic processes are primarily about making molecules. Cells cannot survive without constantly reinventing themselves: making new amino acids for proteins, new lipids for membranes, new nucleic acids so that they can divide. The wheels of the cell can never stop turning while we still live – and turning wheels consume energy.

So the community of the cell is rather like the stereotypical vision of a society during the Industrial Revolution: a culture of manufacture, in which a large part of the workforce is dedicated to generating energy. There are hundreds or even thousands of power plants in every cell of our liver, where energy production is paramount. Like the dark satanic mills of William Blake's *Jerusalem*, they produce waste as well as useful things; but cells have, on the whole, more efficient means of ensuring that they do not foul their own beds.

Since energy production is essential for life, a perusal of the machinations of metabolism can be a worrisome affair. We see what a fragile thing life is – for, if only this or that process were to be disrupted, the whole system would grind to a halt, just as our social structure hangs on the premise of an uninterrupted electricity and gas supply (not to mention clean air and sufficient water). It is a testament to evolution's inventiveness that it has shaped a machine (if you will forgive an Enlightenment metaphor) so robust that it can run for eight decades or so, God willing, before its parts begin irreparably to fail.

Into the fire

But let us begin with some simplicity. Or at least, with the appearance of the same. In the 1850s, the British scientist Michael Faraday delivered at the Royal Institution in London a series of lectures on the 'Chemical History of a Candle'. He wished to show that in the incandescent glow of the candle's flame one could read the whole of chemical science as it was then understood – and more than just chemistry. 'There is no better, there is no more open door', he said, 'by which you can enter into the study of natural philosophy'.

To the trained eye, any natural phenomenon will serve as Blake's grain of sand, a window to the infinite universe. The candle will do, especially in nineteenth-century London, when showcase lectures did not have to be visually stunning. I do not really know the details of how a combustion engine works, but I know that it is not so very different from how a candle burns. It uses oxidation, the combination of some flammable fuel with oxygen in the air to produce heat and, in this case, light.

Even combustion of paraffin wax is still not understood in all its intricate details; certainly, there are many important aspects to the problem that Faraday did not know. Yet the essence of combustion is the essence of all chemical processes that generate energy. First of all, it is a downhill process.

That is the crucial point about chemical change, and about all other processes of change in the universe – they have a downhill and an uphill direction, and naturally enough they proceed in the downhill direction. What determines the topography? In the end, it is that now semi-mystical thing called entropy. Here is the Second Law of Thermodynamics, against which no one may appeal: in all processes of change, the total entropy of the universe increases.*

Popular culture equates entropy with disorder. It is not a bad shorthand: a system's entropy is a measure of how many ways its atoms can be reconfigured without any noticeable difference. If someone were to come into my office and reshuffle the chaos of paper into different heaps, the chances are I would not notice. But if they were to rearrange all the papers in the carefully ordered filing system that I can only dream about, I would soon detect the change. Crudely speaking, ordered systems have fewer indistinguishable configurations – a lower entropy – than disordered ones.

The Second Law is, then, really an expression of the following: it is more likely that a system will progress from a more-ordered to a less-ordered state, simply because there are more of the latter than the former. When one is dealing with systems that contain countless trillions of molecules, this probabilistic statement becomes a near certainty. The Second Law is a law only because its violation is overwhelmingly improbable.

But of course things *do* sometimes get more ordered. Water vapour condenses into symmetrical, six-pointed snowflakes. We can go out and arrange a pile of bricks into a house, and not be arrested for transgression against cosmic law. This is true, but does not violate the principle that entropy must increase. For the Second Law applies to the universe as a whole. Order can be bought at the price of a more-than-compensating increase in chaos elsewhere. Usually this compensation is paid in heat. Yes, we can build a wall by hard manual labour – but by the time we have finished, we will have

---

 * Strictly speaking, this applies to irreversible change. Changes that can be reversed take place with zero change in entropy. By reversible, I do not mean putting something back where you took it from – all the body heat, all the movements of air, all the frictional rubbing, mean that picking up an object is irreversible.

radiated enough body heat to increase the disorderly thermal motions of our surroundings sufficiently for the entropy balance sheet still to be in credit. Order is earned through disorder.

Living cells maintain their organization apparently in the face of entropy's exigencies. This led the physicist Erwin Schrödinger to suppose, in his book *What is Life?*, that the answer lay in the somewhat nebulous concept of 'negative entropy', which living organisms extract from their surroundings. This sounds suspiciously like a kind of vitalism dressed up as thermodynamics; and one still encounters today talk of enzymes as 'anti-entropy devices'. Yet there is nothing special or mysterious about the mechanics of life. Feel that hand (armpit, tongue) again. What are you doing, but pumping out entropy into your environment?

I said that chemical change has a downhill direction – but sometimes it appears to go uphill. Enzymes are especially good at herding molecules uphill, although this can happen in the non-living world too. But in such cases, the reversal of direction is driven by some even more energetic downhill process. You can pull an object up a slope if it is coupled via a pulley to some more massive object moving downhill. The lighter weight rises as the heavier falls. Says chemist Peter Atkins, 'Understanding . . . biochemistry is essentially seeking heavy weights behind subtly hiding screens.'

Energy production in biochemistry is largely about pulling up as many weights as possible before a single heavy weight reaches the end of its fall. For animals like us, this heavy weight is the energy stored in the molecules we eat.

Alternatively, we might draw an analogy with harnessing a river for hydroelectric power. The water is going to go on falling whatever we do – it is never going to rise spontaneously back up the mountainside. The aim, then, is to capture as much of the energy of the cascade as possible. This is one of the major differences between our body's powerhouses and the candle's flame. In the flame, combustion rages uncontrolled, and all we get are heat and light. In the body, combustion takes place in a tightly controlled, graded sequence of steps, and some chemical energy is drawn off and stored at each stage.

Burnt sugar

A power station burns coal, oil, or gas – but it is much more than a flaming grate writ large. Burning is just a means to an end. The heat is used to turn water into steam; the pressure of the steam drives turbines; the turbines spin and send wire coils whirling in the arms of great magnets, which induces an electrical current in the wire. Energy is passed on, from chemical to heat to mechanical to electrical. And every plant has a barrage of regulatory and safety mechanisms. There are manual checks on pressure gauges and on the structural integrity of moving parts. Automatic sensors make the measurements. Failsafe devices avert catastrophic failure.

Energy generation in the cell is every bit as complicated. A brief description cannot hope to do justice to the awesome beauty of the system. The cell seems to have thought of everything, and has protein devices for fine-tuning it all.

The main inputs are fuel and oxygen: food and air. Starve our cells of oxygen and their flame goes out. We might delicately char the fuel before we swallow it, since these subtle beginnings of combustion create compounds that delight our palate. But everything from a chocolate bar to a spinach leaf to a pig's trotter must then be broken down to a more homogeneous fuel. This happens during digestion, when enzymes in the stomach and intestines break apart our culinary creations into their raw molecular ingredients.

There is a variety of energy-rich components in food, which can essentially be classified as either carbohydrates, fats, or proteins. Carbohydrates are polymers formed from molecules of the sugar glucose joined into long chains. Fats (also called lipids) may be broken down into so-called fatty acids and monoglyceride molecules such as glycerol during the digestive process. They yield twice as much energy as the same mass of carbohydrates, and the heart obtains 65 per cent of its energy from them.

Glucose, then, is one of the principal 'heavy weights' of metabolism. Its descent by enzyme-assisted combustion drives the formation of energy-rich molecules called adenosine triphosphate (ATP), which are used to power other processes in the cell. ATP is a biochemical power pack. A great many enzymatic reactions require ATP to drive them uphill.* ATP is the key to the maintenance of cellular integrity and organization, and so the cell puts a great deal of effort into making as much of it as possible from each molecule of glucose that it burns. About 40 per cent of the energy released by the combustion of food is conserved in ATP molecules.

ATP is rich in energy because it is like a coiled spring. It contains three phosphate groups, linked like so many train carriages. Each of these phosphate groups has a negative charge; this means that they repel one another. But because they are joined by chemical bonds, they cannot escape one another's baleful influence. Straining to get away, the phosphates pull an energetically powerful punch.†

The links between phosphates can be snipped in a reaction that involves water, for which reason it is called hydrolysis ('splitting with water'). Each time a bond is hydrolysed, energy is released. Setting free the outermost phosphate converts ATP to adenosine diphosphate (ADP); cleave the second phosphate and it becomes adenosine monophosphate (AMP). Both severances release comparable amounts of energy.

---

  * ATP is not the only power source for the cell's machinery, but it is the most common. Some enzymatic reactions use other, similar energy-rich molecules, particularly guanosine triphosphate (GTP).
  † Biochemical purists will note that this is a simplified argument for why ATP hydrolysis releases energy.

Good digestion

The business of breaking down food begins as soon as it enters the mouth, for saliva contains digestive enzymes called amylases, which start chopping up carbohydrate polymers into glucose. This is why food starts to taste sweeter when it has been chewed. In vegetables the digestible carbohydrate is largely starch; the cellulose of the plant cell walls is also a glucose polymer, but is resistant to the attack of amylases.

In the stomach, the food receives more severe treatment. Here, hydrochloric acid makes the gastric juices about as corrosive as battery acid. The acid loosens up the molecular coils of proteins in the food, ready for destruction by a gastric enzyme called pepsin.

The stomach is really just a storage chamber for undigested food, however. The disassembly process begins in earnest in the small intestine, into which the stomach releases its contents. The intestinal juices are spiced with a cocktail of small, hardy enzymes designed for specialized demolition jobs. While amylases attack carbohydrates, other enzymes break up the food proteins, fats, and nucleic acids. The fragments are absorbed through the intestinal lining and pass into blood and lymph vessels, which distribute these nutrients all around the body. The lining is covered in tiny folds and finger-like protrusions called microvilli, which vastly increase its surface area to ensure that nutrients are absorbed efficiently. Unfolded, the human small intestine would cover a tennis court.

Digestive enzymes are produced in the pancreas, a gland with ducts that empty into the small intestine. But these potent molecules are built to break apart the very molecules from which our cells are made – so how do they not destroy our own tissues?

The enzymes are manufactured with a molecular safety catch attached, which renders them inactive. In this form they are called zymogens. Not until they reach the intestine or stomach is the safety catch (a loop of polypeptide chain) removed, generally by another purpose-built enzyme. The digestive tract of the gut is coated with a layer of mucus, which protects it from digestion enzymes. If the protective coating becomes too thin, the corrosive juices go to work on the exposed tissues, causing an ulcer.

The contents of the stomach are broken down over several hours following a meal. But the body needs fuelling even after digestion is completed. So it stocks up reserves that can be called upon later. About one-tenth of the mass of our liver and 1 per cent of our muscle consists of a substance called glycogen, a compact, highly branched glucose polymer. The liver and muscle cells make glycogen from some of the sugar they receive, and store it in the form of small granules about one- to four-thousandths of a millimetre across – the pantries of the cell.

If the amount of sugar in the bloodstream falls below a certain level, the body knows it is time to start feasting on its glycogen reserves. Low sugar levels trigger the formation of two hormones in the pancreas – glucagon and epinephrine – which tell cells to start breaking apart their glygocen into glucose.

Too much sugar in the blood triggers a different warning system, inducing the formation of a hormone called insulin in the pancreas. Insulin is the signal for cells to start converting glucose to glycogen – to store it rather than burn it. As an indicator of a fuel glut, insulin also signals for cells to focus on manufacturing – for example, making proteins and fats (another emergency fuel reserve) – rather than energy generation.

Insulin is a peptide (protein-like) hormone, and is genetically encoded. A relatively common defect in the genetic message leads to the production of a precursory molecule to insulin (called proinsulin) that a manufacturing enzyme cannot convert to insulin itself in the normal manner. This failure to make insulin is one of the main causes of diabetes, and means that people with the disease have to administer regular doses of the hormone to regulate their blood sugar levels.

Wheels within wheels

Burning sugar is a two-stage process, beginning with its transformation to a molecule called pyruvate in a process known as glycolysis ('sugar-splitting'). This involves a sequence of ten enzyme-catalysed steps. The first five of these split glucose in half in an uphill process, powered by the consumption of ATP molecules: two of them are 'decharged' to ADP for every glucose molecule split. But the conversion of the fragments to pyruvate is a downhill affair that permits ATP to be recouped from ADP. Four ATP molecules are made this way, so that there is an overall gain of two ATP molecules per glucose molecule consumed. Thus glycolysis charges the cell's batteries.

Pyruvate then normally enters the second stage of the combustion process: the citric acid cycle, which requires oxygen. But if oxygen is scarce – that is, under anaerobic conditions – a contingency plan is enacted whereby pyruvate is instead converted to the molecule lactate.* This happens, for instance, if we are exercising so vigorously that the rate of oxygen supply is unable to keep pace with the rate of glycolysis, which the body steps up to meet the high energy demand. As lactate accumulates in muscle tissues it produces the painful symptoms of a 'stitch'.

Anaerobic metabolism is a relatively inefficient way of harvesting the energy of glucose. So extreme exercise leads quite quickly to muscle fatigue: energy is consumed faster than it can be generated, regardless of how much glucose is available. It is this that limits the

---

* There is so much going on in the body that one can rarely afford a general statement, and this is no exception; for some tissues convert substantial amounts of glucose to lactate even under aerobic conditions.

sprinter. The long-distance runner, meanwhile, finds a sustainable pace for which the full process of aerobic (oxygen-driven) metabolism can take its course by way of the citric acid cycle.*

This process is conducted in mitochondria, sausage-shaped compartments distributed many hundred to each human cell (Fig. 21). The first thing a mitochondrion does is convert pyruvate enzymatically to a molecule called acetyl coenzyme A (CoA). The breakdown of fatty acids and glycerides from fats also eventually generates acetyl CoA.

The cycle is a sequence of eight enzyme-catalysed reactions that transform acetyl CoA first to citric acid and then to various other molecules, ending with one called oxaloacetate. This end is a new beginning, for oxaloacetate reacts with acetyl CoA to make citric acid. In some of the steps of the cycle, carbon dioxide is generated as

21. Mitochondria are self-contained compartments in cells. They are responsible for generating energy.

* Animals with higher metabolic rates can sustain exertion for longer periods, because they make ATP faster. The hummingbird can flap its wings furiously almost indefinitely, like an inexhaustible sprinter.

a by-product. It dissolves in the bloodstream and is carried off to the lungs to be exhaled. Thus in effect the carbon in the original glucose molecules is syphoned off into the end product carbon dioxide, completing the combustion process.

Also syphoned off from the cycle are electrons – crudely speaking, the citric acid cycle sends an electrical current to a different part of the mitochondrion. These electrons are used to convert oxygen molecules and positively charged hydrogen ions to water – an energy-releasing process. The energy is captured and used to make ATP in abundance.

The electrons do not flow as if down a metal wire, however; they are carried by a molecule called nicotinamide adenine dinucleotide (NAD). Two of the reactions in the citric acid cycle add an electron and a hydrogen atom to a positively charged ion of NAD, converting it to a molecule denoted NADH. This is the electron vehicle. Transfer of electrons from NADH molecules to an oxygen molecule initiates the formation of water from oxygen and hydrogen, while regenerating NAD. This is fed back into the citric acid cycle (Fig. 22).

There are further wheels within wheels. NADH does not give up its electrons directly to oxygen. Rather, this downhill process is conducted in several stages, and each one is tapped to charge up the mitochrondrion's batteries. The mitochrondrion is bounded by a relatively permeable outer membrane – through which come the raw ingredients that fuel the citric acid cycle – and an impermeable, highly convoluted inner membrane liberally studded with enzyme molecules. Within the inner membrane, NADH relinquishes its electrons to one of these membrane enzymes, whereupon the electrons are passed down a chain of other proteins. Eventually the electrons reach a membrane protein called cytochrome c oxidase, which has a site for binding oxygen molecules. It is this enzyme that finally transfers the electrons to oxygen.

22. **Energy production in the cell from burning sugars takes place in two steps: glycolysis and the citric acid cycle. The first is a linear sequence of reactions; the second, taking place inside the mitochondrion, can be regarded as a series of intermeshing cyclic processes.**

The oxygen-binding site of cytochrome c oxidase will bind certain other small molecules or ions even more strongly than oxygen – for example, cyanide and carbon monoxide. If this happens, the electrons can no longer reach the oxygen, and so this part of the machinery ceases to turn. Jamming of this cog causes the whole of the mitochondrion's mechanism to seize up – the citric acid cycle ceases – and the system is no longer able to produce ATP. The existing stocks exhausted in minutes. So cyanide and carbon monoxide are potent poisons, which can cause rapid death by asphyxiation. The same applies for any substances that disrupt the proteins involved in the chain of electron transport from NADH to cytochrome c oxidase. Such substances include some of the most deadly poisons; this part of the cell's energy-generating machinery is particularly sensitive to attack.

The membrane proteins in the electron-transport chain pass on their bounty downhill to the next in line, and use some of the energy released to pump hydrogen ions from inside to outside the inner membrane. Thus, as the electron hops from one protein to the next, the concentration of hydrogen ions in the space between the inner and outer membrane increases.

This build-up is like the charging of a battery. The difference in hydrogen-ion concentration across the membrane, and the consequent difference in electrical charge of the two regions, is like the difference in electrical potential – the voltage drop – between a battery's terminals. Or you might think of the membrane proteins as pumps driving water back up a mountainside into a reservoir, from which energy can be extracted by letting the water run down again.

The hydrogen-ion reservoir drives ATP synthesis in the mitochondrion by powering a device very much like a miniature water wheel. The hydrogen ions flow back across the inner membrane through a membrane protein called ATP synthase, which harnesses the energy to make ATP from ADP (Fig. 22). ATP synthase has two main components. Its base, firmly lodged in the membrane, is a cylindrical channel through which hydrogen ions can flow. To the end of this channel, which opens onto the inside of the inner membrane, is attached a circular protein structure containing six globular subunits arranged in a ring. As the hydrogen ions pass, this circular head turns, and with each revolution it makes more ATP. ATP synthase is the nub of the cell's energy-generating machinery, and the elucidation of its structure and much of its mechanism of action won Paul Boyer, John Walker, and Jens Skou the Nobel Prize for chemistry in 1999.

Glycolysis and the citric acid cycle are thus strikingly different kinds of metabolic process. One is anaerobic, the other aerobic. Glycolysis is potentially self-contained: one of its steps requires NAD, which is usually made in the citric acid cycle but which can be regenerated anaerobically instead by converting pyruvate to lactate. The two processes look very much like two independent kinds of metabolic pathway bolted together.

This is, in all probability, precisely what they are. The mitochondria are thought to have once been separate bacterial organisms – a notion supported by the fact that they possess their own DNA, distinct from the main genetic library in the cell's nucleus (see page 45). It is believed that, around two billion years ago, the mitochrondria entered a symbiotic (co-dependent) relationship with single-celled organisms that metabolized anaerobically using the glycolytic pathway.

Around this time, the spread of green algae (primitive plant life) led to a sharp increase in the amount of oxygen in the Earth's atmosphere; before then, there was rather little oxygen in the air. The mitochondria-like bacteria were able to 'breathe' oxygen using the citric acid cycle. Symbiosis between the anaerobic cells and the aerobes arose because the anaerobic cells made pyruvate, which the aerobes used for fuel, while the aerobes produced NAD to help drive the anaerobes' glycolytic process. Eventually, the anaerobes engulfed the aerobic organisms to become single, composite oxygen-breathing cells. The rest, as they say, is history.

Yeast never took this step: it is still a glycolytic anaerobe, living off sugar. But instead of turning pyruvate into lactate, yeast cells convert it to ethanol, to the delight of brewers for millennia. This is the process of fermentation, from which stemmed the whole of our understanding of enzyme action. It too generates carbon dioxide, the end product of combustion, which bubbles out of the brewing vats of the world.

Something in the air

Oxygen enters the body through the lungs. But it is not very soluble in blood, and so cannot be carried to the mitochondria simply by dissolving. (Carbon dioxide, in contrast, is soluble enough to make its own way from cells out to the lungs in the bloodstream.) Oxygen is carried by red blood cells packed with the protein molecule haemoglobin, which has a strong avidity for oxygen. Red blood cells could hardly be more single-mindedly designed for their job. Unlike other cells, they contain no DNA or RNA, no compartments and almost no other enzymes – they are basically sacks of haemoglobin. All the genetic and enzymatic machinery required for making this protein is ejected at the last moment as the cells mature.

The red colour of blood is akin to the red of rust, the hallmark of iron. At the heart of the haemoglobin molecule are iron atoms, caught in washer-shaped molecular traps called porphyrins. These iron-loaded porphyrins are known as haeme groups; they absorb blue-green light strongly and so look bright red. Each haemoglobin molecule contains four haeme groups, whose iron atoms provide attachment sites for oxygen molecules.

As an oxygen transporter, haemoglobin must be able to bind its cargo tightly, but also to let it go again. How can it do both? Haemoglobin employs a clever trick that is often used when a protein has to modify its behaviour once it has bound its target molecule. Put simply, one part of the protein can 'feel' what is going on in another part.

In the blood vessels of the lung, where oxygen is plentiful, the attachment of one oxygen molecule to a haeme group sends a shudder through the protein that encourages the other three haeme groups to take up oxygen too. Haemoglobin gives up its oxygen primarily in muscle tissue, where it transfers the molecule to myoglobin, another oxygen-binding protein, which has an even higher affinity for oxygen. Once one oxygen molecule has been taken from haemoglobin, a 'reverse shudder' weakens the grip of the other three haeme groups on their own cargo, and they release the oxygen more readily. These shudders – small but significant changes in the folded shape of the protein chain – are called allosteric motions.

All vertebrate animals and many invertebrates transport oxygen using haemoglobin, although the exact shape of the protein differs in different species. Arthropods and molluscs use a different oxygen-binding protein called haemerythrin, which also contains iron but not bound in a haeme group. This has a violet-pink colour when charged with oxygen, but is colourless when not. Some marine invertebrate creatures use a different metal for transporting oxygen: their oxygen-binding protein, called haemocyanin, is blue, betraying the presence of copper. They are the true blue-bloods of the ocean.

Leaf power

The oxygen-rich atmosphere of Earth is the work of plants. It is a by-product of photosynthesis, the biological use of the sun's energy to make molecules. Since plants and photosynthetic bacteria lie at the base of the food pyramid, all life on Earth is ultimately solar-powered. Without plants we are done for: we are given not our daily bread, and our cattle starve in barren pastures.

Photosynthesis is an old, old process. Algae were photosynthesizing at least three and a half billion years ago, when the continents were newly formed and ungreened. These organisms were the first *autotrophs*: 'self-feeders', making their own molecules from little more than light, water, and carbon in the air. About 60 billion tons of carbon are plucked every year from the carbon dioxide in the atmosphere and are turned into energy-rich biomass. Some of this we eat, burn, arrange into houses and tables, feed to livestock, pulp into paper, spin into cloth. Much of it falls, is broken down by micro-organisms, and is released back into the atmosphere as volatile carbon compounds. Over geological time, some will be buried, compressed into coal, or decomposed into oil or gas.

Photosynthesis depends on molecules that interact with light, absorbing some of its energy and channelling it into chemical processes. While mammalian cells have fuel-burning factories in

23. Plants capture sunlight and convert it to chemical energy in chloroplasts, filled with the stacked sheets of thylakoid membranes.

the form of mitochondria, the solar-power centres in the cells of plant leaves are compartments called chloroplasts (Fig. 23). The process conducted herein is, broadly speaking, the reverse of glucose metabolism. The chloroplast takes carbon dioxide and water, and from them constructs the sugar. 'Burning' glucose is an energetically downhill process, so it follows that the manufacture of glucose in photosynthesis is uphill. This is why the plant needs the energy of light rays to do it. Yet the plant uses this energy not just to create glucose for weaving into the cellulose walls of its cells, but also – and just as importantly – for making ATP molecules to drive the cells' chemistry.

There are several similarities between the processes of aerobic metabolism and photosynthesis. Both consist of two distinct sub-processes with separate evolutionary origins: a linear sequence of reactions coupled to a cyclic sequence that regenerates the molecules they both need. The bridge between glycolysis and the citric acid cycle is the electron-ferrying NAD molecule; the two sub-processes of photosynthesis are bridged by the cycling of an almost identical molecule, NAD phosphate (NADP).

In the first part of photosynthesis, light is used to convert NADP to an electron carrier (NADPH) and to transform ADP to ATP. This is effectively a charging-up process that primes the chloroplast for glucose synthesis. In the second part, ATP and NADPH are used to turn carbon dioxide into sugar, in a cyclic sequence of steps called the Calvin–Benson cycle (Fig. 24).

The first process takes place at the surface of a folded membrane inside the chloroplast, called the thylakoid membrane. This is studded with clusters of molecules called 'photosystems', in which light-absorbing molecules called photopigments initiate

24. **During photosynthesis, captured solar energy is used to split water and to make ATP, which then drives the conversion of carbon dioxide to glucose.**

light-powered reactions. At the heart of the photosystems, called the photosynthetic reaction centre, is a molecule called chlorophyll *a*. This absorbs red and blue light strongly, and is thus responsible for the green colour of leaves.

When chlorophyll receives light energy, it becomes 'excited', like an apple tree that gets shaken. In its excited state, chlorophyll is less able to hold on to its outer electrons, and one of them comes free. The electron is passed on to an enzyme; once the enzyme has gained two electrons from 'shaken' chlorophylls, it can transform a positively charged ion of NADP to NADPH. The electron-deficient chlorophylls are then replenished with electrons plucked from water molecules in another light-powered reaction. The water is broken into hydrogen ions and oxygen atoms. The oxygen atoms combine into two-atom oxygen molecules, which the plant releases through openings in the leaf surface.

The electron taken from water is passed to chlorophyll along a chain of molecules embedded in the thylakoid membrane. Each transfer step is a downhill process that releases energy, some of which is tapped to pump hydrogen ions into the inner space of the thylakoid membrane. This imbalance is then harnessed as an energy source by ATP synthase molecules lodged in the membrane, which perform their windmilling conversion of ADP to ATP.

But the task of photosynthesis is not finished with water-splitting and the production of the energy source ATP and the electron source NADPH. These two ingredients are released into the fluid outside the thylakoid membrane, called the stroma, where they drive the reactions of the Calvin–Benson cycle, which turn carbon dioxide to sugar. These processes are called the 'dark reactions', because they do not require light directly. In 1961 US chemist Melvin Calvin won the Nobel Prize for deducing most of this sequence.

Chemists are currently interested in designing artificial molecular systems that, like chloroplasts, harness sunlight to drive chemical synthesis. A team at Arizona State University has mimicked the chloroplast in synthetic cell-like (and cell-sized) structures called liposomes: hollow, spherical membranes made from lipid molecules. The researchers peppered liposome membranes with designed molecular assemblies that perform the same task as photosynthetic reaction centres – using light energy to pump hydrogen ions into the liposome's hollow interior.

The researchers inserted molecules of ATP synthase into their liposomes, which release the hydrogen ions and make ATP into the bargain. The hope is that the energy stored in ATP can be tapped for chemical synthesis – for example, to conduct some industrially useful biochemical reaction.

Ending with a bang

Glucose and candle wax (a hydrocarbon) are both 'embodied energy': breaking their bonds with oxygen releases energy as heat (unless channelled into other forms). But some chemists seek molecules that pack a bigger punch. How much energy can be crammed into a molecule?

Alfred Nobel occupied himself with that question in the nineteenth century, and the resulting irony – the fortune amassed from the invention of dynamite, which now funds an annual peace prize – is well known. Nobel's innovation was the invention not of an energy-rich molecule but of a means for packaging an existing explosive into a form that was less likely to blow up in one's face.

The oldest explosive is gunpowder, a mixture of sulphur, nitre (potassium nitrate), and charcoal devised in China around the eleventh century AD. This was deployed in the West with little change (and terrible effect) until the nineteenth century, when scientists began to uncover ways of making a bigger bang. In 1845 the Swiss chemist Christian Schönbein discovered nitrocellulose, a compound made by treating cotton fibres (cellulose) with nitric and sulphuric acid. This was the first 'semi-synthetic' polymer, a product both of nature and of the chemist's art. It was later developed into celluloid, a hard plastic, and rayon, the first 'artificial silk'. But nitrocellulose was also explosive, and became known as gun cotton.

Gun cotton's violent nature was hard to control; early attempts to manufacture it led to several deaths. In 1847 the Italian chemist Ascanio Sobrero synthesized a similarly hazardous substance called nitroglycerine. Alfred Nobel began to study this compound in 1859, attempting to find a way of rendering it stable until deliberately detonated. Persisting despite an explosion in 1864 that killed his younger brother, Nobel found that mixing nitroglycerine with a clay called kieselguhr produced a putty-like explosive that could be handled safely. He called it dynamite. In 1875 Nobel introduced gelignite or 'blasting gelatine', a jelly-like mixture of nitroglycerine and gun cotton, which was more powerful than either substance on its own.

These explosives were used largely for mining and construction blasting, and they made Nobel's fortune. But inevitably they were adopted for military use too. The British and French armies of the late nineteenth century used cordite, an explosive similar to blasting gelatine; the German troops took to trinitrotoluene (TNT), which explodes only if detonated with a secondary explosive. In *Brave New World* Aldous Huxley tells us of the rare chemistry of this lethal combination:

$CH_3C_6H_2(NO_2)_3 + Hg(CNO)_2 =$ well, what? An enormous hole in the ground, a pile of masonry, some bits of flesh and mucus, a foot, with the boot still on it, flying through the air and landing, flop, in the middle of the geraniums . . .

All of these compounds are organic substances containing the nitro group: a nitrogen atom with two oxygen atoms attached. The explosiveness of nitro compounds derives from the fact that nitrogen atoms recombine to form nitrogen molecules when the materials are ignited. These molecules have very stable bonds whose formation releases a lot of energy. The oxygen atoms, meanwhile, stimulate the combustion process, allowing it to happen very quickly. Developing more powerful explosives has been largely a matter of finding ways to pack more nitro groups into an organic compound. RDX or cyclonite, an explosive used in today's weaponry, improves on TNT in this respect. The most powerful explosive in current production is a nitrogen-rich compound called HMX, an abbreviation for 'high-melting explosive'.

In early 2000, a potentially still more energy-packed nitro compound was devised by chemists at the University of Chicago. Called octanitrocubane, it consists of a cube of eight carbon atoms, to each of which is attached a nitro group. Not only is this molecule extremely nitro-rich, but the cubic shape means that the bonds between carbon atoms are highly strained and easy to burst open. And the compact molecules should be able to stack together into a very densely packed crystalline form. The initial experiments have not yielded this high-density form, but calculations predict that, if it can be made, it will have an explosive energy content greater, gram for gram, than any known non-nuclear explosive.

Alfred Nobel's prize-giving bequest seems to have been a gesture provoked by the regret he felt for the lethal and destructive applications of his work. Clearly, not all chemists will feel this way about weapons-related research. My own view, for what it is worth, is that such work is arguably a form of scientific misconduct. But, more importantly, work on high explosives demonstrates that scientific research does not split cleanly into an amoral 'pure' science and a dirty and socially accountable 'applied' science or technology. Making octanitrocubane was a feat of technical brilliance; it was also funded by the US Defense Department.

Explosives reveal the Janus face of molecular science. They are part of its fun – for how many budding chemists have never sought out the classic exploding reactions one can conduct in a school lab? I know of at least one respected scientist who was expelled from his school for almost destroying it. (He now studies planet-sterilizing giant meteorite impacts.) Yet it is but a small step from these bangs and flashes to the destruction of Dresden and Hamburg. It is a perilous step.

Chapter 5
Good little movers: molecular motors

After-dinner speeches are not normally notable for launching revolutions. But Richard Feynman, who was engaged to address the West Coast section of the American Physical Society in 1959, was not a normal physicist. One of the most creative scientific minds of the post-war twentieth century, he is most vividly remembered by the world at large as a bongo player, a practical joker, a safe-cracker, the trickster figure of modern science.

Feynman's talk in 1959 was high-spirited but ultimately serious in its intent. He called it 'There's Plenty of Room at the Bottom', and it was about engineering on scales too tiny to see. 'What I want to talk about', he said, 'is the problem of manipulating and controlling things on a small scale'. By 'small', said Feynman, he did not mean 'electric motors that are the size of the nail on your small finger'. He meant small as in atoms.

'Imagine', he went on, 'that we could arrange atoms one by one, just as we want them'. This, he saw, is essentially what the chemist tries to do:

> The chemist does a mysterious thing when he wants to make a molecule. He sees that he has got that ring, so he mixes this and that, and he shakes it, and he fiddles around. And, at the end of a difficult process, he usually does succeed in synthesizing what he wants.

You can see that the physicist's view of what chemists do is hardly more sophisticated than a lay person's. But Feynman's description is really not so different from the one Primo Levi gives when explaining how chemists build molecules as engineers build bridges (see page 24). However, the chemist is traditionally accustomed to regarding his molecule as a substance, something to crystallize and put in a bottle. The physicist, on the other hand, sees it as a construct, like an engine component.

Feynman was essentially wondering whether physicists might figure out how to do what chemists do, but wearing an engineer's hat. Can we build a molecule by pushing atoms into place, one by one? In 1959 such a thing was unthinkable to anyone but a conjuror of the imagination like Feynman.

Yet he was not simply speculating idly. Even at that time, it was clear that technology was getting smaller. The invention of the transistor in the 1940s had shrunk the scale of electronics. Bulky boxes filled with vacuum tubes had been replaced by compact devices containing 'solid-state' circuits made from silicon transistors. The portable transistor radio was on every American beach. Engineers were becoming increasingly skilled at making tiny machine components – much more so, in fact, than Feynman realized. Hoping to provide some small impetus for driving miniaturization technology forward, he offered two prizes of a thousand dollars, to be funded by himself: one for making an electric motor measuring no more than 1/64th of an inch on any side, the other for writing the information from a page of a book in an area scaled down by a factor of 1/25,000. Feynman presumably anticipated that his money would be safe for some years to come – he did not imagine that someone (an engineer named William McLellan) would meet his first challenge within a few months.

Today we can go further. Tiny cogs and motors a tenth of a millimetre across have been carved out of silicon wafers using acid etching or electron beams (Fig. 25). But carving out parts from slabs of material is all very well until you reach a scale of around a tenth of a micrometre: current methods for making integrated circuits in silicon can just about make wires this thin. Beyond that these methods cannot go – it becomes like trying to split a human hair with a bread knife.

Researchers are starting to ask whether this 'top-down' approach still makes sense at such scales. Components this small are closer in size to molecules (medium-sized molecules are a hundred times smaller) than to silicon wafers you can hold and see in your hand. Should we then start making things from the bottom up – from single molecules?

Primo Levi confessed in *The Monkey's Wrench* that chemists fantasize about a tool kit for molecular-scale construction:

we don't have those tweezers we often dream of at night, the way a thirsty man dreams of springs, that would allow us to pick up a segment, hold it firm and straight, and paste it in the right direction on the segment that has already been assembled. If we had those tweezers (and it's possible that, one day, we will), we would have managed to create some lovely things that so far only the Almighty has made, for example, to assemble – perhaps not a frog or a dragonfly – but at least a microbe or the spore of a mold.

Feynman too found inspiration in the molecular devices and artefacts of biology: 'in which chemical forces are used in a repetitious fashion to produce all kinds of weird effects (one of which is the author).' He realized that there are already molecular machines in biology. In 1959 this, had it reached the ears of biologists, would doubtless have been dismissed as the attempt of some foolish physicist to impose his own perspective on a field he clearly knew nothing about. But today biologists are quite happy to talk about proteins as molecular machines.

25. **A micromotor carved from a silicon wafer.**

This chapter looks at some of the most remarkable of these: protein molecules that create movement. They are molecular motors, often called motor proteins. The mini-motor that won Feynman's prize is a clumsy gargantuan in comparison, as a lumbering Diplodocus is to the nimble flea. The biological importance of motor proteins is immeasurable. Without them, we could not move a muscle; no birds would cross the sky, no fish ply the seas. Even bacteria would be immobile. But worse: cells could not divide, so there would be no reproduction. Without molecules to drive movement, there is no life.

Yet for the mechanic of the molecular world, motor proteins say something else. They show that molecular-scale engineering is possible: that we can scale down ideas familiar from the everyday world to the realm of molecules. Motor proteins are not unique in this respect, but they make the point with rare explicitness. I will describe how we might achieve similar goals from scratch, by making our own custom-built molecular motors. This leads us into the arena towards which Richard Feynman's talk was the first clear signpost: the science of *nanotechnology*, which is technology on the scale of nanometres – distances one can measure in molecules.

Front crawl

The shape of a molecule is never fixed: it is always vibrating and waggling its loose, floppy parts. Mechanical motion is ubiquitous in the molecular world.

Yet, in general, either molecular motions are random, like the meandering wriggle of a polymer chain floating in solution, or they average to zero, like the back-and-forth vibrations of a chemical bond. What we need from a genuine motor, in contrast, is motion with a directional bias – what one might call purposeful motion.

Any motor consumes fuel. You could regard this as the inevitable price of orderly motion, a toll imposed by the Second Law of Thermodynamics. Random molecular motion, on the other hand, can be had 'for free' – it is the incoherent molecular jiggle that is heat.

Our bodies conduct many types of directional transport. For example, the motion of cilia – hairlike appendages that line the air passages of our lungs and windpipe – moves a layer of mucus from the lung lining up to our throats, where it accumulates as phlegm. This mucus captures dirt, and so its export keeps the lungs clean. To move the mucus up the windpipe, the cilia cannot just thrash around but have to execute a coordinated sequence of movements, like the arms of a swimmer. They make a whiplike 'power stroke' followed by a slow, crawling 'recovery stroke'. Some single-celled organisms called protists do in fact use cilia on their cell surface to swim through water.

The molecular engine that drives these motions is a protein called dynein. Each cilium contains microtubules (see page 63) arranged around the circumference of a tube called the axoneme. The tubules are joined together in pairs called doublets, like the barrel of a twin-barrelled shotgun. There are nine paired microtubules per axoneme, and they are interconnected by dynein molecules, protruding at regular intervals like the legs of a millipede (Fig. 26).

To move the cilium, the microtubules walk over one another. Each dynein molecule has a 'leg' that bends by undergoing a reaction that consumes ATP. Dynein is basically an enzyme that breaks apart ATP and changes shape as a result. Calcium ions are also needed to

26. **The bending of cilia is driven by dynein molecular motors.**

trigger this reaction. The motion is controlled by nerve signals that trigger the injection of calcium into the cilium.

Since the dynein molecules all point in the same direction, they pull one microtubule doublet over another when they bend. If the molecules were then simply to straighten again, the microtubules would return to their original positions. In order to generate forward motion, each dynein molecule detaches itself from the second microtubule before straightening up, and then reattaches for the next 'power stroke'. Only when its 'foot' is attached to another microtubule can dynein break down ATP and switch to the bent state.

The crawling of doublets one over the other is thus like a ratchet: the cycle of attachment, bending, detachment, and straightening of dynein generates motion in a single direction. But, because the ends of the microtubules are anchored at the base of the axoneme, this sliding of one over another causes bending of the cilium. With proper coordination of the sliding motions, the cilium will bend first this way and then that. The coordination seems to come from a pair of microtubules running down the centre of the axoneme, although exactly how they achieve this is not yet understood.

Dynein plays a broader general role in the world of the cell: it is one of the engines that shuttle objects around. Our cells are laced with an internal rail network of microtubules. From time to time the cell needs to rearrange its compartments, the membrane-branded structures called organelles. Attached to a membrane wall, dynein can pull an organelle along the tracks.

These journeys are one-directional. The ends of microtubules are not equivalent: only from one of them, called the plus end, can tubulin molecules (see page 64) be added or removed. Dynein always moves towards the other extremity – the minus end – of a microtubule, which lies towards the cell's centre. When a cell divides in two, dynein pulls the duplicated sets of chromosomes along the microtubules of the mitotic spindle (see page 65), carrying them towards the centres of the respective nascent daughter cells.

For transport along microtubules in the other (plus) direction, a different motor protein called kinesin is used. Kinesin is perhaps the most anthropomorphic of molecules that induce motion, since it has two 'legs' and executes a waddling 'walk' in comparison with dynein's one-legged inchworm crawl. Kinesin too is powered by a reaction that consumes ATP and alters the protein's shape.

Kinesin is the cell's postman, delivering parcels from one organelle to another. For example, proteins must be sent from their point of manufacture (the endoplasmic reticulum) to the parts of the cell where they are needed. They are packaged inside little membrane spheres called transport vesicles, and kinesin carries a transport vesicle along the microtubule network to the right address.

Muscle power

Our own walking is enabled by muscle contractions and extensions. We are sprung and countersprung with so-called skeletal muscles, which, simply by shortening and relaxing, can control everything from a pianist's delicate finger movements to the pounding of an athlete's thighs.

Skeletal muscle is one of nature's hierarchical molecular materials (see page 53). It is a fibrous composite of bundles within bundles within bundles. Individual muscle cells are extremely elongated and enclose many-stranded cables woven from threads called myofibrils. Within the complex molecular substructure of these strands reside the secrets of muscle contraction.

Seen through the microscope, a myofibril is punctuated by light and dark bands of various widths. These give skeletal muscle a striped appearance at high magnification, which is why it is also known as striated muscle. The sequence of bands repeats periodically along the myofibril strand, and one repeat unit is called a sarcomere. The various bands within a sarcomere are given the kind of anodyne names – A band, H zone, and so forth – that are always telltale indications that no one had the faintest idea, when they were first observed, what they signified.

Noticing in the 1950s that these bands changed width when muscle contracted, Andrew Huxley, Hugh Huxley, and their co-workers proposed the so-called sliding filament theory of muscle action. Their idea was that the myofibril contains toothbrush-like structures facing one another and pushed together so that their bristles interpenetrate. Each sarcomere contains a double set of these pairs of brushes, placed back to back. The dark bands correspond to regions where the bristles interdigitate (creating a high density of molecules), whereas in light bands there is only one set of bristles (Fig. 27). The Huxleys suggested that the myofibril shortens – and so the muscle contracts – by deeper interpenetration of the sarcomere's bristles.

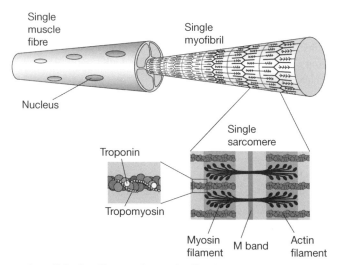

27. **Interdigitating filaments in muscle allow it to contract.**

This movement of filaments over one another is driven by the motor protein myosin, a long thin protein in which two helical chains twist around one another. At either end, the chains terminate in a pear-shaped head. Myosin molecules are gathered into bundles called myosin filaments. Each end of a filament bristles with myosin heads, like a head of corn budding from a sheaf.

Penetrating between the myosin bundles are filaments of a protein called actin. This protein is in fact globular in shape, but the globules link together to form a chain like beads on a necklace. In an actin filament, two chains of actin 'beads' wrap around each other in yet another double helix. The necklace is further adorned by strands of the protein tropomyosin, which wind their way along the actin filament. And at regular intervals sits a globular protein called troponin (see Fig. 27).

Muscle contracts when myosin heads attach themselves to the actin filaments and pull themselves along. The principle is the same as that by which dynein and kinesin move along microtubules: motion is generated by a change in shape of the attached motor protein, driven by the breaking-down of ATP to ADP. The myosin head swings on a hinge connecting it to the rest of the molecule. It kinks, detaches itself from actin, unkinks, and reattaches, and thereby ratchets along the actin filament in a series of power strokes (Fig. 28).

The whole process is under voluntary control, set in motion when a nerve impulse from the brain tells the muscle to tighten or relax. The tropomyosin and troponin proteins on the actin filament provide the switch. Muscles are wired to the brain by nerve cells called motor neurons, which are like wires biochemically 'soldered' to the outside of a muscle fibre. An electrical signal arriving at the end of the motor neuron triggers the release of calcium ions from a web of tubes – the sarcoplasmic reticulum – that extends through the spaces between myofibrils inside the muscle fibre. These calcium ions are captured by

28. Muscle contraction is caused by the movement of myosin molecular motors along filaments of the protein actin.

troponin molecules on the actin filament, prompting the proteins to change shape. This in turn pulls on the tropomyosin strands, which twists the double-helical actin necklace and rotates the actin beads. It is this rotation that exposes the sites on actin to which myosin binds. The entire structure thus contains an elegant molecular transmission mechanism for switching contraction on and off.

Molecular tweezers

Once upon a time molecular scientists had to deduce all they knew about molecules from measurements made on many billions of them simultaneously. This can be a risky business, since we cannot always be sure how such measurements are related to the properties of individual molecules, just as the noise that emanates from a football stadium or theatre hall reveals nothing of the individual conversations people are having. But advances in experimental techniques that enable studies of single molecules – what they look like, how they interact, how they move – have over the past two decades opened up an entirely new realm of molecular studies. We are starting to get to know molecules in person.

One of the critical innovations is the invention of tweezers for molecular manipulation – the very tool that Primo Levi desired. The most remarkable thing about these tweezers is not that they are so fine but that they are literally intangible – they are made of light. They are called optical tweezers, and they trap objects in a very intense light beam. They allow researchers to address the kinds of questions one might ask about everyday mechanical motors: how efficient are they, how much load can they bear, how fast do they move?

Interaction between light and the electrons in molecules can create a force – a kind of 'light pressure' – on an object. If the object is small enough and the light intense enough, the object can be moved by this force. In optical tweezers, the intersection of two or more laser beams sets up a spot of extremely bright light. A small object within this bright pool experiences a light pressure from all sides that prevents it from moving in any direction. It is caught in an optical trap between the tweezers of the laser beams. If the beams are moved, the object is pulled along with them.

The force generated by a single motor protein can be measured by tethering either the motor or the object it moves along (an actin filament, say) to a microscopic plastic bead clamped between optical tweezers. Motion generated by the motor tugs the bead away from the centre of the trap by an amount proportional to the force generated.

Using a bead as a handle, optical tweezers can be used to do extraordinary things with molecules. Kazuhiko Kinosita at Keio University in Japan and his co-workers attached beads to each end of an actin filament and then pulled one end hither and thither until it was threaded through a loop, creating a molecular knot (Fig. 29). They tightened the knot until it broke. Because the actin filament is somewhat stiff, like a sapling branch, it is weakened when sharply curved, and the force required to break the knotted filament was far lower than that needed to pull apart an unknotted filament.

Optical tweezers are not the only tool for handling molecules one at a time. Devices called scanning probe microscopes, devised in the

29. Optical tweezers were used to tie this knot in a strand of actin. The microscopic beads attached to each end acted as 'handles'. The actin is made visible under a microscope by shining light on it to make it fluoresce.

1980s (and used to take the images shown in Fig. 5), have proved immensely valuable not just for observing but for manipulating the molecular world. One of these instruments, called the atomic force microscope (AFM), allows researchers to probe the mechanical properties of molecules – how stiff or stretchy they are, for instance. A molecule can literally be grasped at one end by the AFM and pulled like a piece of elastic.

Motors by design

One of the most prominent prophets of nanotechnology is K. Eric Drexler, an independent scientist who heads the Foresight Institute in California. Drexler's vision, which is built around the idea of molecular-scale robotic assemblers that can put together any molecular machine (including themselves) atom by atom, has been influential on the public perception of nanotechnology's promise (and dangers), but rather less so among scientists. Some scientists worry that Drexler's idea of atom assemblers fails to take account of heat that must inevitably be released when atoms are combined. Moreover, the shapes of molecules are many and varied, but not arbitrary: there is no guarantee that a particular molecular-scale blueprint for a nanotechnological component will correspond to a stable or realizable arrangement of atoms.

Drexler first outlined his ideas in his 1986 book *Engines of Creation*, in which the protagonists (and sometimes antagonists) were nanotechnological robots. But in terms of what is already technologically feasible, a scratch-built, controllable molecular motor would in itself be an engine of creation, however primitive. With such a device, molecular-scale rods, girders, and other construction parts might be shifted into place ready for welding together.

Whereas motor proteins are powered by ATP, some researchers think that synthetic molecular motors could be light-powered. In 1999 a team of chemists led by Ben Feringa at the University of

Groningen in the Netherlands devised a molecular rotary motor, in which a rotor spins in a single direction driven by light. They exploited the process of photoisomerism: the light-induced interconversion of two different forms (isomers) of a molecule, which have the same chemical constitution but different shapes.

They constructed a molecule containing two linked propeller-like units (Fig. 30). Initially the propellers sit on opposite sides of the molecule: the so-called *trans* isomer. But ultraviolet light converts the molecule to the *cis* ('sis') isomer, in which both propellers are on the same side. So as not to bump into one another, the propeller blades twist – one upwards, one downwards. If the molecule is warmed up above 20 °C, the blades switch to the opposite configuration: that which twisted downwards now bends upwards, and vice versa. In this configuration the molecule is slightly more stable. Irradiating it with another dose of ultraviolet light then brings about the reverse switch from the *cis* to the *trans* form. But because of the propeller flip that preceded it, the *trans* form is now subtly different from that at the start: the propellers both bend down rather than up. Heating the molecule to 60 °C restores the original configuration.

The overall result of this four-step process is that one of the propeller blades makes a full revolution with respect to the other, in a predetermined direction. If the molecule is kept above 60 °C and continuously irradiated with ultraviolet light, it will spin smoothly: a light-powered molecular motor.

A different rotary device was made by Ross Kelly and co-workers at Boston College. They constructed a molecule consisting of a three-bladed propeller connected by an axle to a 'brake' that hindered the rotation of the propeller. Without the brake, the propeller rotates – but at random in either direction. The researchers aimed to use the brake to ratchet the propeller in only one direction, by executing a series of chemical reactions between

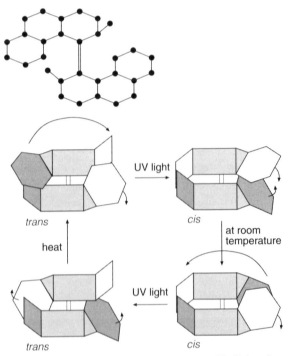

30. **A scratch-built molecular rotary motor powered by light. The top image shows the carbon-atom framework.**

blade and brake. But they have not yet found a way to pull their prop through more than a third of a full turn.

Both of these devices are simplistic and neither can yet be harnessed to perform a useful task. But they show how, in principle, molecular motors might be constructed. The sequence of bond making and breaking required to move Kelly's motor looks cumbersome, but a similar sequence is after all needed to produce linear motion with kinesin and myosin. At the molecular scale, such things can happen quickly enough to give the appearance of smooth motion.

Natural nanotechnology

Synthetic molecular motors have a long way to go before they can compare with natural motor proteins. Does it, then, really make sense to try to build them from scratch, or might one instead adapt motor proteins to nanotechnological ends? Some researchers have isolated motor proteins from the cell and chemically modified them so that they can perform new tasks.

In 1997 Stanislas Leibler at Princeton University and co-workers made devices from the motor protein kinesin that could arrange microtubules into organized patterns. They linked four kinesin molecules together chemically, forming an assembly rather like a creature with four sets of legs. When mixed with microtubules and fed with ATP, these kinesin constructs pulled the tubules one past another until they became organized into star-shaped structures (Fig. 31) very much like those formed in the first stages of cell division (see page 65).

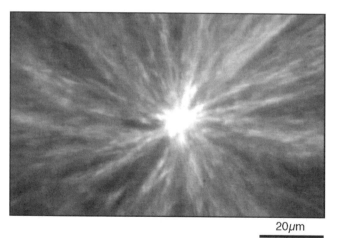

20μm

31. Microtubules organized into star-shaped structures by semi-synthetic molecular motors made from modified proteins.

At the University of Washington in Seattle, Viola Vogel and co-workers have used kinesin to propel microtubules over surfaces in a selected direction. They attached kinesin molecules to a surface coated with the polymer polytetrafluoroethylene (PTFE), better known as the non-stick coating Teflon. The PTFE was applied by rubbing a block of it over the surface, whereupon the polymer film acquires striated grooves and ridges in the direction of rubbing. The polymer chains are thought to be aligned with these ridges. Kinesin molecules become attached preferentially on the ridges, which means that they form oriented rows. These rows act as linear tracks along which microtubules can be passed: the kinesin molecules pass the tubules to one another like a bucket brigade. In cells it is the kinesin molecules that are mobile and the tubules that are 'fixed'. But in these experiments the motor proteins are tethered to the surface, so their walking motions propel the microtubules instead.

The most dramatic and exciting amalgamation so far of biomolecular motors with artificial microengineering was described towards the end of 2000 by Carlo Montemagno and co-workers at Cornell University in Ithaca, New York. They commandeered a molecular *rotary* motor to turn a tiny metal propeller about 150 nanometres wide and nearly ten times as long. The enzyme ATP synthase, we saw earlier, has a head that spins on a membrane-bound spindle as it performs its task of converting ADP to ATP (see page 83). Montemagno and colleagues fixed this head to the top of a microscopic pedestal etched from nickel metal, and then they attached the metal propeller to the spindle. Under the right conditions, ATP synthase can work in reverse, breaking down ATP to ADP and spinning as it does so. The researchers initiated this process by feeding their rotors with ATP, and saw them revolve under the microscope at around five revolutions per second (Fig. 32).

Studies like this raise the exciting prospect of using molecular motors to move molecules around in a controlled way – something that brings a whole new dimension to synthesis at the molecular

32. A microscopic metal propeller attached to the spindle of the rotary motor protein ATP synthase (*a*) rotates when the motor is driven with ATP. *b* shows an array of many such constructs. Only those for which the propellers appear here as non-vertical are working as intended: the synthesis is not yet successful in every case.

scale. No longer would chemists have to rely on the random wandering and chance encounter of molecules floating in solution: they could instead guide them precisely where they are meant to go. Because nature has already devised a wondrous array of molecular machines for such purposes, I suspect that molecular nanotechnologists will increasingly make use of the cell's machinery rather than trying to design devices from first principles. This applies not just to the generation of mechanical motion but to areas such as energy generation, sensors, and information processing. We may then see a fusion of biology with disciplines once regarded as quite different, such as mechanical and electronic engineering. Because the union will bring about results that none of the fields can achieve on its own, we could call it *biosynergetic engineering*.

Chapter 6

Delivering the message: molecular communication

Each of us is a new world. The molecular view of life reveals that the appropriate analogy for a cell is a city, teeming with molecular inhabitants. Our many-celled bodies are thus collaborations between communities. One cell communicates with another as London does with Liverpool, New York with Philadelphia: messages are passed down wires or carried from place to place by courier. Goods are transported hither and thither along the transportation network of the blood and lymph circulatory systems. It is just as Berzelius said: 'This *power to live* [is] the result of the mutual operation of the instruments and rudiments on one another.'

Molecular biology has long been content merely to document the social webs of the cell: deducing which molecules speak to which, and how they come and go. But ultimately this is not enough. We need to know also what is said, and how the messages are passed on one to another. This is information that a pharmaceutical chemist can use to develop drugs. The fundamental challenge for medical science is to learn how to take part in the body's molecular conversations: to intercept harmful or unpleasant messages, to send out new warning messages, to prevent undesirable interactions.

It is largely as a result of these efforts in the field of biochemistry that chemistry itself is undergoing something of a reinvention. The imagination of chemists is fired up by what they see to be possible in biology. Although much of the chemicals industry is devoted to the manufacture of 'passive' products – new plastics, cement, glue, paint, synthetic fibres – drug molecules were always a little different. For their task is to partake in a dynamic process, to engage in the active life of the cell. They are like actors primed for a dramatic role; indeed, they often play their part by impersonation. Now chemists are beginning to appreciate that this kind of dynamism can be achieved in purely synthetic chemical systems too. And so chemistry is becoming less about the properties of individual molecules and more about how groups of different molecules behave together – forging and breaking relationships, modifying each other's tendencies, sending out signals. Chemistry is becoming a science of *process*.

This is the attitude that underpins much of the work I discuss in this book: the development of molecular solar cells, of chemical sensors, of a molecular nanotechnology, and of molecular devices that process information. Much of the research in this area is gathered under the umbrella of *supramolecular chemistry*, which means chemistry beyond the molecule: the science of molecules in communication.

In this chapter I shall explore a few of the ways in which molecules communicate in biology, before providing a glimpse of how a similar gregariousness can be encouraged in synthetic molecules. As ever, we must remember that, while nature is inspirational, it is also parsimonious and blind. Biology uses a limited range of materials, and has a tendency to adapt a good solution endlessly for new purposes rather than exploring a completely new avenue each time. Just as the Jumbo jet is not a scaled-up pigeon, so the wise molecular engineer takes from nature principles but not blueprints.

Molecular mail

Italy and Germany became countries when their patchworks of small kingdoms and city states agreed to respect a central authority – to collaborate towards a greater good for all. The body cannot behave as a coherent entity unless cells do likewise. This means that there must be mechanisms for sending commands, edicts, and calls to action throughout the entire realm. Nerve signals from the brain are one means by which the body coordinates its actions. They are the body's telephone system.

But general messages sent *globally* throughout the body are posted like a mass mail shot into the bloodstream, in the form of molecules called *hormones*. These are diverse both in form and in function. Some hormones are large proteins; others are small organic molecules. Some are soluble in water, others insoluble (which means that courier molecules are required to carry them through the bloodstream). Some convey urgent, immediate messages such as 'Run away!' Others have long-term effects, promoting growth or the development of sexual characteristics.

All hormones are the product of the *endocrine system*, a series of glands that constitutes the overarching regulatory system for the whole body (Fig. 33). We have seen already how the hormones insulin and glucagon, produced in the pancreas, control the blood's sugar content (see page 78). Similarly, the rate of metabolic processes in cells is regulated by the hormones thyroxine and triiodothyronine released from the thyroid gland. These hormones affect energy production and oxygen consumption, in part by altering the heart-beat rate.

The control centre of the endocrine system is the hypothalamus, a gland in the brain. The hypothalamus is connected to the pituitary gland, which sits just below it, from where hormones are dispatched

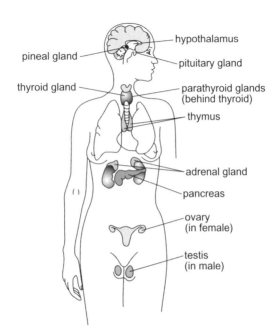

33. The body's endocrine system, a series of hormone-producing glands.

to other glands. For example, a fall in metabolic rate prompts the hypothalamus to send a molecule called thryotropin-releasing hormone to the pituitary. This gland in turn starts to send out thyroid-stimulating hormone, which triggers the thyroid gland into action.

All of the hormones released from the pituitary are peptides: small protein-like molecules. Antidiuretic hormone, for instance, controls the body's water content by regulating the production of urine in the kidneys. Growth hormone provides a stimulus for cell multiplication, and plays a key role during childhood and adolescence. It also stimulates localized growth of tissue when repairs need to be made – for example, during wound healing.

The adrenal glands manufacture some important steroid hormones. These are insoluble molecules with a carbon backbone consisting of several small rings joined together. Some steroids, such as cortisol, regulate the storage and use of the body's energy resources: the conversion of glucose to glycogen and the breakdown of proteins into amino acids. Bodybuilders and athletes use these hormones (legally or otherwise) to build up body mass and muscle.

As you might anticipate, the hormone adrenaline is also a product of the adrenal glands. Along with noradrenaline, it is released rapidly into the bloodstream in response to stress. Both hormones quicken the heart rate and dilate the blood vessels, increasing the oxygen supply to muscles so that they are prepared for extreme exertion.

The sex glands – ovaries in women, testes in men – release the hormones that differentiate the sexes and trigger changes in growth during puberty. Testosterone stimulates sperm production in men. Oestrogen and progesterone control the female menstrual cycle; their production is regulated by hormones released from the pituitary gland, called follicle stimulating hormone (FSH) and luteinizing hormone (LH).

These two hormones regulate ovulation during the menstrual cycle. In the early days of pregnancy, high levels of oestrogen and progesterone in the bloodstream inhibit the production of FSH and LH and suppress ovulation. Birth-control pills have the same effect: containing oestrogen and progesterone, they persuade the woman's body that she is already pregnant.

Production of oestrogen declines when a woman is in her thirties, and particularly during the menopause. Side effects of low oestrogen levels include an increased susceptibility to coronary heart disease and to bone loss, which are two of the prime motivations for administering oestrogen in hormone replacement therapy. The treatment remains controversial, because long-term doses of oestrogen can have unwelcome side effects of their own, including an enhanced susceptibility to breast cancer and different forms of heart disease.

Switched on

How is a hormonal message read? This depends on the nature of the message. Some hormones can be posted right through the walls of cells, wherein they bind to some receptor protein. This activates the receptor to stimulate the transcription of a particular gene, making a protein that the cell needs. This so-called direct gene mechanism of hormone action works for hormones that are small and insoluble, so that they can penetrate the fatty cell membrane.

But many hormones, particularly those comprised of peptide and protein molecules, get no further than knocking on the doors of the cell. They are received by butlers at the cell surface – receptor proteins whose job it is to convey the message to others inside the cell.

Like most other molecular communication, the passing of a message from hormone to receptor protein is an intimate affair. Molecules show no inhibitions: they speak to one another through close embraces. Lacking any other means of recognition, molecules identify one another by 'touch', through binding events in which the receptor latches onto a target (substrate) with precisely the right shape, like a key fitting in a lock. Each hormone-receptor protein on a cell surface has a binding site sculpted to fit around the hormone.

Despite the variety of messages that hormones convey, the mechanism by which the signal is passed from a receptor protein at the cell surface to the cell's interior is the same in almost all cases. It involves a sequence of molecular interactions in which molecules transform one another down a relay chain. In cell biology this is called *signal transduction*. At the same time as relaying the message, these interactions amplify the signal so that the docking of a single hormone molecule to a receptor creates a big response inside the cell.

It works like this. The receptor proteins span the entire width of the membrane; the hormone-binding site protrudes on the outer surface, while the base of the receptor emerges from the inner surface (Fig. 34). When the receptor binds its target hormone, a shape change is transmitted to the lower face of the protein, which enables it to act as an enzyme.

The process that the enzyme catalyses is the 'activation' of a so-called G protein, attached to the inner surface of the membrane. G protein is short for guanine-nucleotide-binding protein: the protein holds onto a molecule of guanosine diphosphate (GDP). When a hormone-charged receptor interacts with a GDP-laden G protein, the G protein first replaces the GDP with guanosine triphosphate (GTP, analogous to energy-rich ATP), and then breaks in two. The half that binds the GTP becomes an enzyme, and travels off to activate another enzyme at the inner surface of the cell wall. Commonly, this other is adenylate cyclase, a protein that converts ATP into cyclic AMP (cAMP).

The participants of all these processes are stuck to the cell wall. But cAMP floats freely in the cell's cytoplasm, and is able to carry the signal into the cell interior. It is called a 'second messenger', since it is the agent that relays the signal of the 'first messenger' (the hormone) into the community of the cell. Cyclic AMP becomes attached to protein molecules called protein kinases, whereupon they in turn become activated as enzymes. Most protein kinases switch other enzymes on and off by attaching phosphate groups to them – a reaction called phosphorylation. The action of a protein kinase initiates a cascade of reactions, since each activated kinase can act on several enzyme molecules, each of which in turn can do its job many times. In this way docking of one hormone to its receptor can affect many molecules inside the cell: the signal becomes amplified.

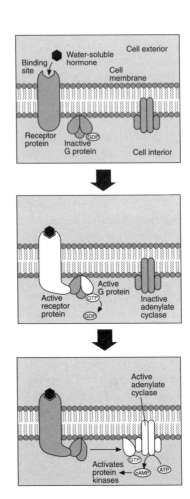

34. How G proteins work.

The process might sound rather complicated, but it is really nothing more than a molecular relay. The signal is passed from the hormone to its receptor, then to the G protein, on to an enzyme and thence to the second messenger, and further on to a protein kinase, and so forth.

The G-protein mechanism of signal transduction was discovered in the 1970s by Alfred Gilman and Martin Rodbell, for which they received the 1994 Nobel Prize for medicine. It represents one of the most widespread means of getting a message across a cell membrane. Some hormones attenuate rather than stimulate cell processes; in such cases the activated G proteins might exert an inhibitory effect on their target enzymes rather than activating them. In other cases the second messenger might be a small molecule other than cAMP: certain activated G proteins trigger the release of calcium ions from the calcium-binding protein calmodulin, for instance.

And it is not just hormonal signalling that makes use of the G-protein mechanism. Our senses of vision and smell, which also involve the transmission of signals, employ the same switching process. The roof of the nasal cavity is lined with smell sensors called olfactory hairs, which are attached to the ends of nerve cells that carry signals to the olfactory bulb – the 'smell centre' of the brain. The cell walls of the olfactory hairs are studded with receptor proteins designed to bind particular odorant molecules that enter the nose.

There are several hundred different kinds of odorant receptors, each one with a binding site shaped to accommodate a specific common odorant. We can, however, discriminate between a wider range of odours than this, because each odour is typically the result of a complex blend of different odorant molecules. The olfactory bulb forms an 'image' of the smell from the mixture of impulses that it receives from different receptors, much as we recognize a person's face from the sum of the different component parts.

In smell signalling, the cAMP produced by G proteins binds to a membrane protein called a sodium channel, whereupon the channel opens up and lets sodium ions flow into the cell. This triggers a nerve impulse, which passes to the olfactory bulb. The same basic process generates visual signals in the optic nerve when stimulated by light.

Our sense of taste is due largely to our olfactory system. The taste buds in our tongues can distinguish only relatively crude signifiers of taste: sweetness, bitterness, saltiness, and sourness. The full delight of a matured cheese or freshly baked bread comes mostly from the odorant molecules they release.

All in the mind

Hormones can trigger complex webs of biochemical action, but the messages they bear are pretty crude, related to the exigencies of growth and survival. It is quite another matter that communication between molecules gave birth to the Sistine Chapel, to *The Magic Flute*, to the theory of relativity. Yet the mind is, after all, made of molecules.

At the same time, the mind is still a mystery – one of the great remaining mysteries of science. Some scientists argue that the mind will never be able fully to comprehend itself, that the self-referential nature of the problem will always create blind spots. Others believe that a scientific explanation of consciousness is on the horizon. Either way, it is likely that the secrets of the mind lie far beyond the molecular realm, embedded in questions about the behaviour of complex, highly connected information networks. Here we see the limitations of reductionism – for the molecular processes of thought are now fairly well mapped out, yet their collective consequences are barely sketched.

The brain contains somewhere in the region of a billion to a hundred billion brain cells or *neurons*. That is nothing to write home about – other organs are comparably populous. But the distinguishing feature of the brain is the complexity of the communication network between these cells. Each neuron makes around a thousand links, so there may be up to a hundred trillion interconnections in the brain – about the same as the number of stars in a thousand galaxies like our own. On such a transportation network, you would be lost in an instant. The degree of connectivity in the integrated circuits of computers is nowhere near so great, and it is no surprise that computers, for all their literal-minded speed, fail miserably at some tasks that a child can do in a flash.

Neurons send nerve signals – in essence, electrical pulses – to one another along tubular channels called axons. The axon ends in a series of branches whose tips push up against the membranes of other neurons. At these junctions, called synapses, a nerve signal is transmitted from one neuron to another. Neurons also sprout many shorter, bushy branches called dendrites, which collect information from the axons of other cells. The axons are, if you like, the motorways of the brain, stretching from one neuronal city to the next. They end in slip roads that connect up at synapses to the city road system of the dendrites.

Although axon signals are electrical, they differ from those in the metal wires of electronic circuitry. The axon is basically a tubular cell membrane decorated along its length with channels that let sodium and potassium ions in and out. Some of these ion channels are permanently open; others are 'gated', opening or closing in response to electrical signals. And some are not really channels at all but pumps, which actively transport sodium ions out of the cell and potassium ions in. These sodium-potassium pumps can move ions 'uphill' – from regions of low to high concentration – because they are powered by ATP.

In its 'resting' state, the axon has an imbalance of sodium and potassium ions inside and outside that sets up a charge difference, or voltage, across the membrane: the fluid inside has a small

35. **Electrical pulses are sent down the axon by the opening and closing of ion channels.**

negative charge (the 'resting potential') relative to that outside. When a signal is sent down the axon, some of the gated sodium channels open up, altering the distribution of ions and reversing the imbalance: the inside becomes positively charged relative to the outside. This region of reversed voltage opens up sodium channels ahead of it, so that it moves along the axon. At the same time, the channels behind close up and the resting potential is restored. In this way, a voltage pulse or 'action potential' travels down the axon (Fig. 35).

At a synapse, this nerve impulse is transmitted from the axon to another neuron. The signal is generally first converted from electrical to chemical form. A small molecular messenger called a neurotransmitter conveys the signal across the space (called the synaptic cleft) between the terminal membrane of the axon and the membrane of the other neuron. The neurotransmitter is packaged up inside a bubble-like membrane that merges with the axon's cell wall, releasing the molecular message into the synaptic cleft. It travels to the outer surface of the other neuronal membrane, where it becomes bound to receptor proteins.

A diverse array of molecules serve as neurotransmitters. Some are simple amino acids – glycine and glutamate – or molecules derived from them, such as serotonin and dopamine. The molecule acetylcholine is a neurotransmitter that carries messages between the central nervous system and muscle cells at neuromuscular junctions (see page 103). When acetylcholine binds to its receptor on a muscle cell, the receptor is transformed to an open sodium channel. Sodium ions rush into the cell, changing the voltage across the cell membrane and opening up a voltage-controlled calcium channel. This triggers a rise in calcium ion concentration inside the cell, which stimulates muscle contraction.

Acetylcholine illustrates the general function of a neurotransmitter: to open or close an ion channel, thereby altering the voltage across the membrane in which the receptor sits. This converts a chemical message back to an electrical one. Acetylcholine does the job directly, since its receptor is itself an ion channel. Some other neurotransmission pathways function rather differently: they use a second messenger to transfer the message from the neurotransmitter to an ion channel, again with the mediation of G proteins.

Is it surprising that the G-protein signal transduction mechanism appears in so many different contexts? Not really. As the complexity of multicelled organisms evolved, cells with ever more specialized functions arose from common ancestors with more general functions. Tried-and-tested mechanisms for certain tasks would be retained, though adapted where necessary. That, after all, is why we share genes with yeast and bacteria. The G-protein pathway is an effective way of passing a chemical message from the outside to the inside of a membrane, and amplifying it in the process. The cell's motto is: if it works, find a way to use it.

Pleasure and pain

Neurotransmission is a common target for drugs – beneficial, harmful, pleasurable, or, depending on the dose, all three. The nervous system is one of the most vulnerable parts of the body: if nerve impulses are blocked, we cannot move. Many animals make toxins that cause paralysis in their prey by attacking the sodium-potassium pumps or the voltage-gated ion channels in axons, blocking the progress of action potentials.

Muscle action is also affected by drug molecules that resemble acetylcholine and so compete with it in binding to the receptor proteins at neuromuscular junctions. Nicotine, the active ingredient of tobacco, is one such: it binds to a certain class of acetylcholine receptors in muscle and causes the associated stimulatory sensations: increase in heart rate and dilated pupils. Why the sensation is pleasurable is not, however, fully understood. Curare is a lethal toxin present in the bark of a South American plant, which was once extracted and used by the indigenous people to poison arrow tips. Curare binds to the same class of acetylcholine receptors as nicotine, but does not activate them – so muscle action is prevented. An animal poisoned with curare will die of asphyxiation, unable to inflate its lungs. The medieval poison hemlock works in the same manner.

Whereas some neurotransmitters stimulate neurons, the role of others is to quieten them: to suppress the firing of action potentials. These are said to be inhibitory, and include glycine and the molecule gamma-aminobutyric acid (GABA). Our thoughts are a complex interplay of stimulation and inhibition, as neurons weigh up the various signals they receive from their neighbours and decide whether or not, on balance, they should fire off a salvo themselves.

Hallucinogenic drugs such as LSD (lysergic acid diethylamide) and mescaline overexcite the brain by enhancing the stimulatory effects of serotonin. The poison strychnine blocks inhibitory signals, leading to uncontrollable muscle spasms and a particularly unpleasant death. Depressants assist the binding of inhibitory neurotransmitters or (like alcohol) interfere with the action of excitory neurotransmitters.

Drugs that relieve pain typically engage with inhibitory receptors. Morphine, the main active ingredient of opium, binds to so-called opioid receptors in the spinal cord, which inhibit the transmission of pain signals to the brain. There are also opioid receptors in the brain itself, which is why morphine and related opiate drugs have a mental as well as a somatic effect. These receptors in the brain are the binding sites of peptide molecules called endorphins, which the brain produces in response to pain. Some of these are themselves extremely powerful painkillers.

Cannabinoids, the active ingredients of cannabis, also bind to inhibitory neuroreceptors in the brain to produce pain relief. The natural target of these receptors is a molecule called anandamide, which, like endorphins, is produced in response to pain signals. A closely related molecule called oleamide seems to be the biochemical trigger that induces natural sleep.

Not all pain-relieving drugs (analgesics) work by blocking the pain signal. Some prevent the signal from ever being sent. Pain signals are initiated by peptides called prostaglandins, which are manufactured and released by distressed cells. Aspirin (acetylsalicylic acid) latches onto and inhibits one of the enzymes responsible for prostaglandin synthesis, cutting off the cry of pain at its source. Unfortunately, prostaglandins are also responsible for making the mucus that protects the stomach lining (see page 78), so one of the side effects of aspirin is the risk of ulcer formation.

One of the surprising recent discoveries in neuroscience was that extremely small inorganic molecules can also act as neurotransmitters. Carbon monoxide and nitric oxide – both of them two-atom molecules – serve this function. They are both poisonous in large doses, because they compete with oxygen in binding to haemoglobin. But 'the poison is in the dose', and in small amounts nitric oxide does some important things. It triggers the dilation of blood vessels, which can relieve stress on the heart. This is why nitroglycerin, which decomposes to release nitric oxide, is administered to treat heart problems. The improvement in circulation initiated by nitric oxide provides the basis for the drug Viagra, which is used to treat erectile dysfunction in men.

Supramolecular chemistry

In recent decades, scientists have become interested in mimicking, in synthetic systems, some of the molecular communication processes of the cell. There are many motivations for this. Drug development is often a matter of concocting a good disguise, so that a synthetic molecule will pass itself off as a natural one and bind preferentially to a receptor, blocking or initiating a biochemical signal. Signal transduction in the eye's retinal cells and in the olfactory system suggests the concept of molecular sensors that can detect light or other molecules with high sensitivity. Molecular engineers are looking to the olfactory apparatus for inspiration in designing 'artificial noses' that can identify complex mixtures of molecular components.

The principle loudly extolled by nature is the 'lock and key': molecules get together when one fits with the other.* To turn such a 'recognition' event into a communication process, the binding event should trigger some change in the receptor that allows it to relay the signal downstream. In biology this relay process is commonly catalytic: binding turns the receptor into an active enzyme. But the

* This metaphor was first used by the German chemist Emil Fischer in 1894 to explain how enzymes are so selective about the transformations they catalyse.

signal might also be passed on in other ways: by the emission of light or release of an electron, for instance, or (as in the case of the acetylcholine receptor) by the creation of an electrochemical potential.

Building artificial signal-transduction processes at the molecular scale is a common objective in supramolecular chemistry. From its outset, this discipline was biologically inspired. In the 1960s, the French chemist Jean-Marie Lehn investigated so-called crown ether molecules that would recognize and bind specific metal ions. Lehn was interested in transporting ions such as sodium and potassium across lipid membranes. Although this can be mediated by protein channels and pumps, another strategy is to engulf the ion within a molecule that will 'dissolve' in the fatty interior of the membrane wall. Such molecules exist in nature, and are called ionophores. A typical example is valinomycin, a ring-shaped peptide with a central hole into which a potassium ion will fit. Crown ethers are synthetic mimics of valinomycin: they too are ring-shaped, and will bind a metal in their central cavity. The metal ion is held more or less securely depending on the relative sizes of the ion and the hole. If the hole is too big, the metal 'rattles around' and is only loosely bound; too small, and it will not fit. So crown ethers can be tailored to fit specific metal ions – to display molecular recognition, in other words.

By the 1970s, Lehn and others were making synthetic receptor molecules of all shapes and sizes, with cavities designed to accommodate a wide range of inorganic and organic targets. These 'guest' molecules are held in place by interactions with their hosts that are weak relative to the covalent bonds that hold the atoms together in the molecules themselves. In this way the guests can be picked up and released again. That is how valinomycin works as a metal-ion transporter: it captures the ions on one side of the membrane and lets them go again at the other side. Supramolecular chemistry is essentially about bringing molecules together into loose associations that can be disassembled back into their separate components.

When crown ethers take up a metal ion, they change shape. Alone, they are rather loose, floppy rings, like rubber bands. With a metal ion in their core, they become organized into relatively rigid structures in which the ring contains zigzag-like kinks: a crown, in other words (Fig. 36). Shape changes of this sort are common when a receptor binds its target.

If binding alone is the objective, a big shape change is not terribly desirable, since the internal rearrangements of the receptor make heavy weather of the binding event and may make it harder to achieve. This is why many supramolecular hosts are designed so that they are 'pre-organized' to receive their guests, minimizing the shape change caused by binding.

But if the idea is to use binding as the trigger for relaying some signal, then a shape change is often crucial. A particularly dramatic change in shape induced by host–guest binding was reported in 2000 by Ulrich Koert and colleagues at the Humboldt University in

○ Hydrogen

● Carbon

○ Oxygen

36. A crown ether is a cyclic molecule that captures a metal ion in its central cavity.

37. An artificial transducer molecule that converts binding of a zinc ion into a signalling event, by changing shape and altering its fluorescence properties.

Berlin. They constructed a receptor molecule that can be considered as a series of 'modules': two arms, two legs, and a 'transducer' unit like a flexible torso connecting the arms to the legs. When the two arms close around a zinc ion, the transducer unit flips and pulls the two legs far apart (Fig. 37). The legs are tipped with fluorescent groups, which change their emission wavelength – from green to ultraviolet – when the distance between them increases. The researchers pointed out how this molecular receptor shows some of the features of a protein receptor involved in signal transduction, responding to binding of the target at one end by altering its shape, and thus its behaviour, at the other end.

A shape change that alters a molecule's light-emitting properties has been engineered in several other synthetic receptors. But using recognition and binding to switch a molecule's *catalytic* behaviour, as in the G-protein signalling mechanism, is rather more challenging, since this means ensuring that the final shape is exactly what is needed for the catalyst to do its job. Harder still would be the task of organizing several molecules into a relay that would carry a message downstream. All the same, the skill of supramolecular chemists is increasing daily, and it would not be at all surprising if we were soon to see artificially devised molecular communication systems that approach the sophistication of those the body uses to run its realm harmoniously.

Delivering the message

Chapter 7
The chemical computer: molecular information

In the end, we are left with the same question: what is life? It will not be answered – or at least, not here. But Schrödinger's answer – negative entropy (see page 74) – for all its shortcomings contains a grain of truth. For it is a necessary but not sufficient characteristic of life that it imposes order on chaos. Chaos is death. If cells cannot send and receive clear messages, if they do things at the wrong time, if their membranes lose their organization, if proteins fail to fold, then life cannot be sustained. We are islands of order in a wild world.

Where does this order come from? Organization can arise spontaneously in inorganic matter too: think of the serried ranks of mare's-tail clouds, or the regular ripples in wind-blown sand. It seems highly likely that this kind of 'self-organization', which can arise unbidden in systems fed with energy that prevents them from achieving a static equilibrium, has a part to play in life's orderliness. But that is not enough. The kind of coordination needed for a cell to copy its chromosomes and divide in two, to make a functioning and reproducible protein molecule, or indeed to grow from a single fertilized egg into a multicelled Mozart, cannot rely on the 'blind' patterning processes that paint the skies and the deserts. There needs to be a firmer hand on the wheel.

We have all heard of what this guiding hand consists. It is DNA, a string of molecular beads chopped and bundled into our forty-six little X-shaped chromosomes. The human genome – our full complement of DNA – is often called the 'book of life'. As I write, scientists have just finished decoding the first draft of this book: they have sketched out in broad detail the molecular messages in each of the chromosomes.

Incautious things are said about the project to map the human genome. One hears, for instance, that a sufficiently skilful engineer could make a human from the information therein. This is nonsense. The body is full of molecules that are not encoded in the genome – it encrypts only proteins, and even those in somewhat garbled and incomplete form. The genome tells us nothing about the lipids that make up cell membranes, let alone about how they are driven by physical forces to aggregate into sheets, loops, and spheres. The genes will not tell us how neural signalling works, how the brain encodes thoughts and sensations in delicately timed trains of electrical pulses. There is no gene for bone, for tooth enamel. The genome is the book of the cell in much the same way as the dictionary is the book of a performance of *Waiting for Godot*. It is all in there, but you will not deduce one from the other.

Nevertheless, the genome *is* an instruction booklet in molecular form. It tells us how to make proteins, the molecules that orchestrate the spectacular molecular performance of life. In this chapter I want to say a little more about the nature of that script: how it is read and enacted. But my ultimate intention is broader. For the molecular scientist, genetics says something truly dramatic and profound about molecules: they can carry and transmit information. The theoretical biologist John Hopfield of Princeton University points out that this is one of the many ways in which biology provides an 'existence theorem' to inspire chemists. 'Mathematicians use this term', he says, 'in reference to a proof that some function which they want to construct actually does exist, and is not impossible. In this sense, the observation of birds flying provides an existence theorem that an engineer should be able to design a flying machine.'

In the same way, genetics shows that it is possible to perform computing with molecules. For computing is nothing but the storage, transmission, and processing of information, all of which the genetic machinery can do. Says Jean-Marie Lehn, 'there is a "molecular logic of living organisms"'.

This is really a corollary of the previous chapter, where we saw that molecules can communicate with one another. A genuine molecular logic is more precise: it requires not simply that one molecule affects the behaviour of another, but that they can transfer and manipulate encoded information in well-defined ways. This is how a computer works, by passing data between switches and memory devices made from semiconducting and magnetic materials.

Computing with molecules is just one of the ways in which the dimension of information is entering molecular science. More generally, chemists are becoming accustomed to the idea that molecules can be programmed to behave in certain ways: that their properties can be written into the fabric of the molecule just as a set of instructions can be programmed into a robot. Lehn says 'The outlook . . . is toward a general science of informed matter.' Such a chemistry is a genuinely new science, in many ways quite different from the traditional chemistry that makes useful substances. It is a science that is about an active 'becoming' rather than a passive 'being'. It is already happening, and we do not know where it might take us.

How the cell 'becomes'

Every book is written in a particular language, and the genome is no different. The language of the genes is a simple code, whose characters are the four nucleotide molecules that represent the beads of DNA's molecular necklace (see page 42). Each of these molecules contains a so-called base, which encodes the information. There are four DNA bases: adenine, cytosine, guanine, and thymine (A, C, G, and T). As DNA is a linear polymer of nucleotide units, the information it uncodes can be represented as a linear string of these four characters. Part of it might look something like this:

GTGGATTGACATGATAGAAGCACTCTACTATATTC

A four-letter alphabet might seem a rather limited system for writing complex messages. But, if we think of this sequence as a code rather than strictly as an alphabet, it can be as complex as you like. We could, for example, denote every letter in the Roman alphabet by a series of several bases: GTG for 'a', GAT for 'b', and so on. The number of permutations of four characters in groups of three is sixty-four – more than enough to encode all the alphabet. Using such a code, we could write the Bible as a string of A's, G's, C's, and T's.

The cell does not have much use for the message of the Bible; what it needs are messages for making proteins. The way that a protein chain folds up is determined by its amino-acid sequence (see page 41) – so the 'information' for making a protein is uniquely specified by this sequence. DNA encodes this information using a cipher just like that suggested above: groups of three bases represent each amino acid. This is the genetic code.*

* I should point out that here I have made one of the many necessary simplifications in descriptions of genetics. We have seen that some of the most important parts of proteins, such as the haeme unit of haemoglobin, do not consist of amino acids but of other chemical groups. These so-called prosthetic groups are added to the protein chain after it has been manufactured, and are constructed by other enzymes. The bare protein chain, without prosthetic groups, is called an apoprotein. Generally it is completely useless until the frills have been added. Truly to deduce the structure and form of a protein, we need to know not only its amino-acid sequence – and thus the sequence of the gene that encodes it – but also the identity and function of proteins that operate postnatally, so to speak, on the apoprotein.

How a particular protein sequence determines the way its chain folds is not yet fully understood. This means that we cannot deduce a gene's function from its sequence alone (although we can sometimes make good guesses). The first draft of the human genome is full of genes of unknown purpose.

Nevertheless, the principle of information flow in the cell is clear. DNA is a manual of information about proteins. We can think of each chromosome as a separate chapter, each gene as a word in that chapter (they are very long words!), and each sequential group of three bases in the gene as a character in the word. Proteins are translations of the words into another language, whose characters are amino acids. In general, only when the genetic language is translated can we understand what it means.

DNA is a double-stranded polymer: two chains are twisted around one another in a double helix. Each of these strands consists of a string of nucleotides, embodying encoded information. But the strands are not identical. They are stuck together by a kind of zip of hydrogen bonds (see page 41) between the bases on one strand and those on the other. Although all the bases can form hydrogen bonds, they have distinct preferences in their pairing relationships: A sticks to T, and G to C. So the twin helices of DNA have complementary sequences: wherever an A appears in one strand, a T appears in the other, and so forth. This means that each gene is written in two versions, echoing each other in a kind of mirror language.

The particular partnerships of bases are dictated by their shapes. The A and G bases are similar molecules, as are C and T. So an A–T pairing has much the same overall shape and size as a G–C pairing. These pairs point inwards between the two coiled strands like rungs on a spiral staircase. Because the rungs are the same size, the strands coil evenly. An A paired with a G would create a bulge, and the resulting distortion would destabilize the pairing. Likewise, if

C and T were to pair up, there would be a constriction in the double helix. A–C and G–T pairings, meanwhile, are discouraged by the disposition of hydrogen-bonding groups in these base pairs. So the pairing preferences are a matter of a good fit – of *complementarity* – between the two partners.

This is one of the crucial points about biological information flow: the transmission of data occurs via molecular-recognition processes, which ensure that each part of the message is read correctly.

DNA is replicated – the genome is copied – when a cell divides. Because the two strands are complementary, each of them can serve as a template on which the other can be put together. If A always binds preferentially to T and so forth, the sequence of a 'naked' single strand will guide unlinked nucleotides to line up in the right order for forming the complementary strand.

In order to act as a template, a single strand is unzipped from its partner by special enzymes. Complementary strands are then pieced together along both of the exposed strands; an enzyme called DNA polymerase catalyses the addition of each new nucleotide. So the two new double helices each contain one strand from the original.

Although enzymes help the process along, the essential information needed for the copying process is already written into the DNA templates. In the early 1980s, Leslie Orgel at the Salk Institute in California and co-workers showed that individual nucleotides can be assembled into polymers on a template of complementary nucelotides without assistance from enzymes. A string of eight C-bearing RNA nucleotides, for example, will act as a template for joining together eight G-bearing nucleotides. Orgel had to cheat a little, however, by using G nucleotides that had been 'activated' by the addition of a reactive chemical group, which helped them join together.

This template-assisted polymerization is not, in itself, replication: the new strand is complementary to the template, not identical. The first example of genuine artificial molecular replication was reported in 1986 by German chemist Günter von Kiedrowski. He used the same templating process, but chose a template that was self-complementary: it was its own complement. The template was a six-nucleotide DNA molecule with the sequence CCGCGG. Its complementary sequence is the same as this, because the two strands of the helix line up with one oriented in the reverse direction to the other. Von Kiedrowski assembled this complement from two three-nucleotide fragments, again activated to help them link up (Fig. 38).

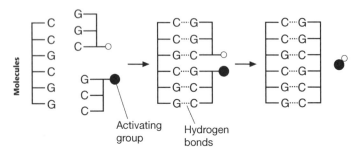

38. **Molecular replication in a six-unit nucleic acid.**

Errors and junk

At some stage during the production of this book I will have received from the publishers page proofs – rough versions of the final pages, prepared from the manuscript I provided. It will (I hope) be a more or less faithful transcription of what I have written. But almost without doubt there will be a scattering of small errors owing to typing mistakes or file-reading glitches. No writer is surprised by such things, because it is inevitable that copying a long and complex message introduces a few mistakes.

The same is true in the genetic processes of transcription (where DNA is copied into RNA) and translation (where RNA is copied into a protein sequence). Occasionally a wrong nucleotide or amino acid will be inserted into the chain, for molecules are not capable of perfect recognition all the time. Probably about one in every twenty or so proteins is incorrectly made.

Does this matter? On the whole, no. It will be an unusual situation if, between me and the publishers, we spot all the typographical errors in this book before it goes to print. But hopefully none of them will be so serious that you are unable to catch my meaning. Similarly, in proteins most of the chain is scaffolding, which holds in place the few amino-acid residues that conduct the protein's catalytic task. An error here and there in the scaffolding might not be serious. Sometimes the consequence of an error might be a completely non-functional molecule; but because the cell makes not one but typically dozens or hundreds of enzyme molecules for any particular task, one or two duds do not matter.

Here I am talking about *random* errors. Far more serious are *systematic* errors, which arise further back in the stream of biological information flow – closer to the ultimate repository of information. A wrongly transcribed RNA molecule could generate hundreds of faulty proteins. So there are enzymes that check the transcription process quite carefully for copying errors, reducing their frequency to around one in ten thousand.

Even errors in transcription are seldom a matter of serious consequence: after all, RNA molecules are ephemeral; the cell can always make more of them. But an error in DNA can be bad news indeed, because there is no way of correcting it once it is in place. One misplaced nucleotide in a gene means that all the RNAs made from that gene, and all the proteins made from those RNAs, contain the analogous fault. Worse still, every cell stemming from division of the genetically faulty cell inherits the same flaw. If the genetic defect occurs in a gamete – a sperm or egg cell – then it is transferred to progeny derived from that gamete. This is why DNA replication is scrutinized extremely carefully by 'proof-reading' enzymes, which permit no more than one error in a billion bases to creep through. Without these molecular proof-readers, we would acquire about 1,000 defective genes in every new cell.

Inheritable errors, the result of mistakes made in DNA replication during production of the gametes, are known as mutations. Once established, they are passed on from parent to offspring right down the genealogical tree. Mutations are responsible for genetically related disorders such as cystic fibrosis, as well as for genetically linked predispositions to conditions such as cancer and heart disease. But, in spite of such terrible consequences, mutations are also the spice of life. Indeed, we owe to them our very existence. If the primitive single-celled organisms replicating in the steamy broth of the early Earth had not picked up occasional mutations – if they had invariably copied their DNA without a single mistake – there would have been no evolution, no emergence of complex life.

Something else will doubtless have happened to my text when the proofs come back from the publishers. Every so often there will be words I did not write. They will not be errors, however, but will make perfect sense: they will be changes made by the editor, and will, I am sure, make the text far easier to read and understand than was my original.

It came as something of a surprise in the mid-1970s to find that genes too need editing. The RNA transcript that peels off from the DNA template is not fit for translation to proteins, for it contains a lot of useless information. These 'primary RNA transcripts' are rather like sentences that have fragments of other sentences inserted seemingly at random. The RNA molecules need heavy editing before they present a clear message fit for translation.

The useless inserts are called introns, and sometimes they make up most of a gene. They are also known as non-coding sequences, since they do not encode parts of proteins. Enzymes snip out the introns from the RNA primary transcript, and splice together the two ends of the coding regions (called exons).

This is just one of the ways in which the supposed 'book of the cell' is mostly garbled nonsense – or boring repetition. It is thought that only about 2–3 per cent of the entire human genome codes for proteins. Some sequences get repeated for good reason. Each human chromosome ends in the sequence TTAGGG repeated about 2,500 times. These sections, called telomeres, are thought to keep the chromosomes stable. They get shortened each time a cell divides, and their eventual erosion contributes to the ageing process. But many other repeat sequences serve no useful function. Transposons are repeats that jump about in the genome, leaving copies as they go. They are thought to be a genetic parasite living in the very core of our being, whose only purpose is to replicate themselves. Introns may be the remnants of ancient transposons that lost the ability to move on.

The protein machinery for cutting, splicing, replicating, and synthesizing nucleic acids provides the principal tools for genetic biotechnology, the manipulation of genomes. Restriction enzymes, for instance, are proteins that can recognize a specific short sequence of DNA and cut the chain at that point. Ligases join loose ends of DNA together. Stretches of DNA can be replicated indefinitely in a test tube using DNA polymerases. The double strands are separated by heating, exposing them for templated replication. Repeated cycles of replication and heating multiply the DNA exponentially. This process, called the polymerase chain reaction (PCR), uses a DNA polymerase taken from a bacterium that lives in hot springs. The enzymes of this bacterium have evolved to withstand high temperatures, and so this DNA polymerase is not destroyed by the heating cycles.

These tools allow scientists to 'rewrite the book': to insert new genes into an organism's genome. Crop scientists are interested in developing plants with genes that confer resistance to pests, or to drought, or to particular herbicides, as well as incorporating genes that improve the flavour of the crop, or its growth rate, or whatever. One potential danger is that genes for herbicide resistance, for example, might become transferred from agricultural crops into weeds, generating new breeds of 'superweed'. The likelihood of this 'horizontal' trans-species transfer of genes is not known.

Some people object to genetic engineering on the grounds that it is ethically wrong to tamper with the fundamental material of life – DNA – whether it is in bacteria, humans, tomatoes, or sheep. One can understand such objections, and it would be arrogant to dismiss them as unscientific. Nevertheless, they do sit uneasily with what we now know about the molecular basis of life. The idea that our genetic make-up is sacrosanct looks hard to sustain once we appreciate how contingent, not to say arbitrary, that make-up is. Our genomes are mostly parasite-riddled junk, full of the detritus of over three billion years of evolution. There seems little that is admirable or elegant in this unruly library; rather, the admiration should be reserved for the cohorts of diligent proteins that painstakingly sift snippets of meaning from reams of nonsense. It is truly amazing how well the whole affair works; but, like most of life, it is a makeshift compromise in which efficiency and tidiness count for little.

Construction blueprints

DNA is a supreme example of 'informed matter'. It is programmed to assemble in a highly specific manner, each nucleotide marrying up with its complement over thousands of base pairs. This kind of programmed self-assembly represents one of the goals of supramolecular chemistry. Atoms do not in themselves display many powers of discrimination; but by making the molecule, rather than the atom, the fundamental building block, supramolecular chemists are able to programme much more guiding information into their bricks and mortar.

Yet DNA provides more than an existence theorem for programmed self-assembly. It can supply the very fabric. Why not use the principles of complementary base pairing to join together DNA girders into structures much more complex than the double helices of the cell?

This concept has been explored by Nadrian Seeman at New York University. He commandeers the enzymatic apparatus of biotechnology to cut and splice DNA into remarkable edifices, such as cagelike polyhedra: a cube and a truncated octahedron (Fig. 39). The edges of these structures are DNA double helices, but at the corners three coils meet in a triple junction. These junctions are cunningly woven: the twin strands go separate ways along different edges, where they intertwine with new strands. Seeman and his co-workers make these junctions by piecing together synthetic DNA with carefully planned sequences.

The trick is then to assemble triple junctions into a three-dimensional geometric molecular object. Seeman gives the branches 'sticky ends' where one strand extends beyond the other.

39. A polyhedral molecular assembly made from double-stranded DNA.

Here the exposed, unpaired bases are ready to match up with those on another strand – but only if it has a complementary sequence. In this way, the ends are selectively sticky and can be assigned one to another, so that the whole structure is programmed to build itself from its component parts. Once the sticky ends are married up to their partners by base-to-base hydrogen bonding, ligation enzymes forge strong bonds to secure the backbone.

For the present time, these molecular constructions are follies of virtuosity: demonstrations of the astonishing control that molecular recognition can provide over nanometre-scale architecture. But Seeman suggests that his DNA frameworks might act as scaffolds for assembling other molecules and materials in useful ways. For example, it is possible to coat DNA strands with silver, transforming them into electrically conducting molecular wires. Might we one day build tiny electronic circuits by 'genetically' programming the wires to link up in a specified pattern?

Moreover, sticky-ended DNA can clip together molecular-scale objects in a selective manner. Chad Mirkin and colleagues at Northwestern University in Illinois have used this idea to assemble little particles of gold into clusters. The particles are just a few nanometres in size – so-called nanocrystals. Each has a tag consisting of single-stranded DNA, but, because the tag sequences are not complementary, the particles remain separate. By adding lone DNA strands whose two ends complement the sequences of the tags, the researchers are able to link the nanoparticles together. The resulting clusters scatter blue light strongly, so the solution turns the colour of wine. Mirkin and colleagues are now developing this technique commercially as a method for simple visual detection of DNA strands with a particular sequence – something that is commonly required in genetic analysis.

By assembling nanocrystals of metals or semiconductors into organized arrays, some researchers hope to be able to build electronic devices far smaller than those currently made with the conventional microfabrication techniques used for creating silicon chips. Semiconductor nanocrystals could act as memory elements for storing electronic information, and they interact with light in ways that could be useful for making light-based information-processing devices. Programming the assembly of nanocrystals using DNA linkers might provide one way of arranging them into circuit patterns. Another possibility has been suggested by the work of Angela Belcher of the University of Texas and co-workers, who used the molecular-recognition properties of proteins rather than DNA. They developed small peptide molecules that would recognize and stick to the surfaces of different kinds of semiconductor. The peptides could 'feel' how the atoms were organized at the surface of the semiconductor crystals. Genetically engineered motor proteins (see Chapter 5) with peptide arms that recognize certain kinds of semiconductor might one day be used to drag nanocrystals around on a molecular building site and arrange them in a circuit pattern.

Molecular logic

Since the invention of the computer in the 1940s, the computing power of new machines has roughly doubled every eighteen months. This trend, known as Moore's law after Gordon Moore, the co-founder of Intel who first pointed it out in 1965, is driven by miniaturization. Computer power increases as it becomes possible to pack more circuit components into a given space. But, if Moore's law is to hold fast for another twenty-five years or so, electronic devices must shrink to nanometre sizes: the scale of molecules.

No one yet knows how that will be achieved, for at such scales the silicon transistor – the workhorse of integrated circuits – becomes too leaky a switch. In order to continue making computers faster and more powerful, a growing school of thought says that their components will have to be individual molecules. This is so different a vision from conventional information technology that it would be a bold or reckless speculator who invests in it.

Yet it is not a new idea. In 1974 US chemists Mark Ratner and Ari Aviram proposed a design for a single-molecule rectifier (a device that passes current in only one direction). Just a few years later, carbon-based polymers were discovered that can conduct electricity, and researchers began to hope that individual molecules of these materials might supply the wires for a 'molecular computer'. The field of *molecular electronics* was born.

But for the next decade or so, nothing much happened. It was an idea before its time, lacking any experimental means to synthesize, arrange, or probe the kinds of molecular devices it dreamed of. In recent years, several paths have converged to create a resurgence in the field, and at last molecular electronics – and its corollary, molecular computing – are starting to gather serious attention from the people who matter: the companies who build computers.

One of the central ingredients of a molecular information-processing technology is the switch: a device to supplant the transistor. In the most rudimentary terms, a switch can exist in two different stable states – 'on' and 'off'. The transistor passes a current when 'on', and blocks it when 'off'. But a switch is useful for information processing only if it can be hooked up to others, so that the devices can talk to each other and pass information to and fro. This is hard to achieve with molecules.

Yet something of the kind was announced in 1999 by James Heath and co-workers from the University of California at Los Angeles, in collaboration with scientists from the computer giants Hewlett-Packard. They interconnected several switches based on an organic molecule to produce an electrically controlled logic gate.

In computer circuits, information is encoded in binary form – as a series of 1's and 0's. A signal of 1 corresponds to a electrical pulse at a certain voltage; a signal of 0 corresponds to zero voltage. Those are the only two signals dispatched through the circuit – there are no signals of ½ or 2. Data is encoded in the sequence of 1's and 0's just as DNA encodes information in a sequence of nucleotide bases. The binary code is simpler than the genetic code: it has only two characters. Each unit of information in the coded signal – each 1 and 0 – is called a binary digit, or bit.

Computers manipulate binary information and perform calculations using logic gates: devices or circuits that make decisions. A logic gate receives one or more input signals and releases one or more output signals. The outputs depend on what the inputs say. An AND gate, for instance, receives two input bits and produces one output. If both inputs are 1's, the output is also 1; but any other combination of inputs generates an output of 0. Simple combinations of logic gates like these can perform arithmetic: for example, reading two numbers encoded in binary form and generating an output that encodes their sum (addition) or their difference (subtraction).

Heath and colleagues constructed AND gates from a molecular switch called a rotaxane. This is an assembly of two molecules: a ring threaded on a rod. The ring is prevented from falling off by big capping units fixed to the ends of the rod. The rod is designed to attract the ring, so that the two molecules interlock spontaneously when mixed together. The end caps are added afterwards. Heath's colleague Fraser Stoddart at UCLA developed techniques for making molecular assemblies like this while at the University of Sheffield in England in the late 1980s.

The researchers arranged rotaxane molecules in a layer on a metal electrode, and deposited fine metal wires on top of them. By applying a voltage to the wires, the molecules could be switched from a low-conductivity to a high-conductivity state. Many thousands of molecules, attached to a single wire, constituted a single, switchable device. The researchers connected several of these devices together to make an AND gate.

In principle, they said, it should be possible to build each device from just a single switchable molecule. But it is difficult to make electrical connections to, and measure tiny currents through, single molecules. Yet even this is not impossible. Mark Reed, James Tour, and co-workers in the USA have measured the electrical conductivity of a single 'molecular wire' connecting two gold electrodes.

Fraser Stoddart, in collaboration with Vincenzo Balzani at Bologna and their co-workers, has demonstrated a different logic operation, called XOR, in a single-molecule gate. Like AND, an XOR gate has two inputs and one output. The output signal is 1 if the inputs are different (0,1 or 1,0) and 0 if they are the same (0,0 or 1,1). The researchers observed this behaviour in a pseudorotaxane – a threaded-ring molecular assembly with no end stoppers, so that the ring can slip off the rod.

Rather than using electrical signals, the device received chemical inputs and produced an optical output. That is to say, it altered its light-emission behaviour (fluorescence) depending on whether two 'chemical signals' were 'on' or 'off' – whether the two chemicals were present or not (Fig. 40). This is analogous to the way that cell-surface receptor proteins work (see page 119) – they send out some kind of signal depending on whether or not they have bound their target molecules.

Using similar principles, A. Prasanna de Silva and Nathan McClenaghan at the University of Belfast in Northern Ireland have combined two molecular logic gates in a way that permits them to conduct elementary arithmetic. They have, in other words, been able to use molecules to count and to perform simple sums, such as $1 + 1 = 2$.

It is, of course, a long way from adding one plus one to making a computer that compares with silicon-based devices. But studies such as these demonstrate an important principle: molecules can be

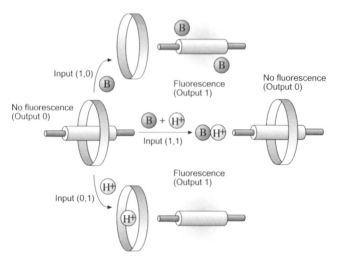

Input (1,0)

No fluorescence
(Output 0)

B

Fluorescence
(Output 1)

No fluorescence
(Output 0)

B + H⁺ → B H⁺

Input (1,1)

Input (0,1)

Fluorescence
(Output 1)

H⁺

40. Molecular logic conducted by a needle-and-thread molecular assembly called a rotaxane.

used for computation, and at the level of one device per molecule. The molecular computer is at last starting to look like more than canny advertising.

The closer we look at the idea, the more we can see similarities with the challenges that the body faces: how to arrange molecules where you want them, how to transmit and amplify signals, how to grow wires between two switching devices (such as neurons), how to cope with errors, how to control the relative timing of events. The computer engineers of the future may need to know a lot of biology.

DNA computing

As if to drive that message home, in recent years some scientists have shown that computing can be conducted using DNA. This brings us full circle, for I began by suggesting that DNA provides a kind of proof-of-principle for molecular computation. But in the cell it provides the programme for making proteins. No one dreamed, until Leonard Adleman suggested it in 1994, that DNA could be used to solve the same kinds of problems as computers. Adleman realized that the genetic code can be used, just like the binary code of computer science, to encode mathematical problems. He showed that biotechnological techniques for manipulating and rearranging DNA can be used to generate all possible answers to such a problem, each one encoded in a molecule of DNA. Techniques for analysing DNA sequences are then employed to screen through all these possible answers and identify the correct one.

For certain kinds of mathematical problem, computers have no short cuts to the right answer. They simply have to test out all options, and select the best. If the number of possible answers is large, this search can take a very long time. Such problems are some of the hardest to solve using conventional computers. A classic example is the 'travelling salesman' problem, which entails working out the shortest route connecting a large number of points in space ('cities') so that each is visited just once.

Adleman showed that, by shuffling and splicing short segments of DNA at random, all the solutions to these problems may be encoded in a test tube of single-stranded DNA molecules. The number of such solutions might be huge – but the number of molecules in a test tube is greater still. And, because all the possible answers are produced and tested at once, rather than one at a time, in principle DNA computing can find the 'best' answer rapidly.

Whether or not DNA computing proves to be useful in a practical sense, it has a strong allegorical appeal. It drives home the message that the molecular basis of life is rooted in the manipulation of information. It is often said that each age tends to interpret the world through models derived from its most advanced technology, and so maybe in the Age of Information we should be wary of becoming too dogmatic about such a (partial) answer to the perennial question that haunted Haldane, Schrödinger, and countless others. It is perhaps more important that we regard this as a demonstration of the fabulously dynamic, interactive world inhabited, unseen and too often unsung, by molecules.

Notes

1. Engineers of the Invisible

1 'The sergeant beckoned the waitress'. F. O'Brien, *The Dalkey Archive* (Normal, Ill.: Dalkey Archive Press, 1993), 80.

2 'Did you ever study the Mollycule Theory'. Ibid. 81.
'Mollycules is a very intricate theorem'. Ibid. 81–2.

4 Primo Levi's famous book is *The Periodic Table* (London: Abacus, 1986).
The quotations in the box are from ibid. 3, 58, 171; the translation is copyright Shocken Books Inc. (1975).

6 John Horgan's book is *The End of Science* (London: Little, Brown & Co., 1997).
The several other recent books I refer to include the (otherwise excellent) M. Kaku, *Visions* (New York: Doubleday, 1997), and J. Maddox, *What Remains to be Discovered* (London: Macmillan, 1998).

7 Roald Hoffmann's book is *The Same and Not the Same* (New York: Columbia University Press, 1995).

8 'a "low" chemistry, almost culinary'. P. Levi and T. Regge, *Conversations* (London: I. B. Tauris & Co., 1989).
Thomas Pynchon, *Gravity's Rainbow* (London: Penguin, 1987).

9 The quotation in the box is from ibid. 249–50.

13 My book is *H_2O: A Biography of Water* (London: Weidenfeld & Nicolson, 1999).

15 'the profession I studied in school'. P. Levi, *The Monkey's Wrench* (London: Penguin, 1987), 142–3.

24 Primo Levi 'put it this way' in ibid. 144.

26 For the story of the aniline dye industry, see A. S. Travis, *The Rainbow Makers* (Bethlehem, Pa.: Lehigh University Press, 1993), and S. Garfield, *Mauve* (London: Faber & Faber, 2000).

29 'as you can imagine'. Levi, *The Monkey's Wrench*, 144.

30 The papers reporting the synthesis of taxol are K. C. Nicolaou *et al.*, 'Total Synthesis of Taxol', *Nature*, 367 (1994), 630; R. A. Holton *et al.*, "First Total Synthesis of Taxol. 1: Functionalization of the B Ring', *Journal of the American Chemical Society*, 116 (1994), 1597; R. A. Holton *et al.*, 'First Total Synthesis of Taxol. 2: Completion of the C and D Rings', *Journal of the American Chemical Society*, 116 (1994), 1599.

32 Müller's group describe their 'big wheels' in A. Müller *et al.*, 'Formation of a Ring-Shaped Reduced "Metal Oxide" with the Simple Composition $[(MoO_3)_{176}(H_2O)_{80}H_{32}]$', *Angewandte Chemie International Edition*, 37 (1988), 1220.

2. Vital Signs

33 'I am not going to answer this question'. J. B. S. Haldane, *What Is Life?* (Welwyn: Alcuin Press, 1949), 58.
'Nature is what we know'. Emily Dickinson (*c.*1863), *The Complete Poems*, ed. T. H. Johnson (London: Faber & Faber, 1975), 332.
'life is a pattern of chemical processes'. Haldane, *What Is Life?*, 61–2.

35 'the cause of most phenomena'. J. J. Berzelius (1812), *Ofversigt af djurkemiens framsteg och narvarande tillstånd: A Report to the Swedish Academy of Science, Presented on 13 August 1810*, translated into English by G. Brunnmark as *A View of the Progress and Present State of Animal Chemistry* (London, 1818), 4–8.

36 'This *power to live*'. Ibid.

45 Allan Wilson's group reported their findings in R. L. Cann, M. Stoneking, and A. C. Wilson, 'Mitochondrial DNA and Human Evolution', *Nature*, 325 (1987)/31. This article was much criticized on methodological grounds: see H. Gee, *Nature*, 355 (1992), 583.

48 'Advances in directed evolution'. J. W. Szostak, D. P. Bartel, and
     P. L. Luisi, 'Synthesizing Life', *Nature*, 409 (2001), 387.

3. Take the Strain

51 Space tethers are described in R. L. Forward and R. P. Hoyt,
    'Space Tethers', *Scientific American*, 280/2 (February 1999), 86.
    For the broader picture on molecular materials, see P. Ball, *Made
    to Measure: New Materials for the 21ˢᵗ Century* (Princeton:
    Princeton University Press, 1997).

56 The many varieties of spider silk are discussed in F. Vollrath,
    'Spider Webs and Silks', *Scientific American* (March 1992), 70.

63 Making materials from artificial genes is described in D. A. Tirrell,
    M. J. Fournier, and T. L. Mason, 'Genetic Engineering of
    Polymeric Materials', *MRS Bulletin* (July 1991), 23.

66 The discovery of carbon nanotubes was reported in S. Iijima,
    'Helical Microtubules of Graphitic Carbon', *Nature*, 354
    (1991), 56.
    For the story of the discovery of the fullerenes, see J. Baggott,
    Perfect Symmetry (Oxford: Oxford University Press, 1994).

69 The tentative step towards the production of nanotube fibres is
    described in B. Vigolo *et al.*, 'Macroscopic Fibers and Ribbons of
    Oriented Carbon Nanotubes', *Science*, 290 (2000), 1331.

4. The Burning Issue

73 The Second Law of Thermodynamics is discussed in depth in
    P. W. Atkins, *The Second Law* (New York: W. H. Freeman & Co.,
    1984).

74 Erwin Schrödinger, *What Is Life?* (Cambridge: Cambridge
    University Press, 1944).

90 The work on artificial mimics of photosynthesis is described in
    G. Steinberg-Yfrach *et al.*, 'Conversion of Light Energy to Proton
    Potential in Liposomes by Artificial Photosynthetic Reaction
    Centres', *Nature*, 385 (1997), 239; G. Steinberg-Yfrach *et al.*,
    'Light-Driven Production of ATP Catalysed by F_0F_1–ATP Synthase
    in an Artificial Photosynthetic Membrane', *Nature*, 392 (1998),
    479.

91 'CH$_3$C$_6$H$_2$(NO$_2$)$_3$ + Hg(CNO)$_2$'. A. Huxley, *Brave New World* (London: Granada, 1977), 48.

92 The synthesis of explosive octanitrocubane is described in M.-X. Zhang, P. E. Eaton, and R. Gilardi, 'Hepta- and Octanitrocubanes', *Angewandte Chemie International Edition*, 39 (2000), 401.

5. Good Little Movers

94 Richard Feynman's article 'There's Plenty of Room at the Bottom' appeared in *Engineering and Science* magazine (February 1960). More accessible versions are available in B. C. Crandall and J. Lewis (eds.), *Nanotechnology: Research and Perspectives* (Cambridge, Mass.: MIT Press, 1992), or at *http://www.zyvex.com/nanotech/feynman.html*.

96 'we don't have those tweezers'. P. Levi, *The Monkey's Wrench* (London: Penguin, 1987), 144.

98 I wish I could point to a good general reference source for nanotechnology, but that is surprisingly difficult. K. Eric Drexler has done most to popularize the subject, albeit in a manner that owes as much to science fiction as to current science – see his *Engines of Creation* (London: Fourth Estate, 1990), and also K. E. Drexler, C. Peterson, and G. Pergamit (1994), *Unbounding the Future* (Reading, Mass.: Addison-Wesley, 1994). And, for critical comment on Drexler's vision of nanotechnology, see G. Stix, 'Trends in Nanotechnology: Waiting for Breakthroughs', *Scientific American* (April 1996), 94; and D. E. H. Jones, 'Technical Boundless Optimism', *Nature*, 374 (1995), 835. Ed Regis's *Nano!* (New York: Bantam Books, 1995), is a readable account, but centres around Drexler's vision. The volume by Crandall and Lewis (above) has some nice contributions, if slightly technical and now outdated. Michael Gross's *Travels to the Nanoworld* (New York: Plenum, 1999), is well worth reading but sometimes idiosyncratic in its choices. The web site of Drexler's Foresight Institute (*http://www.foresight.org*) does a good job of collating up-to-date popular and technical articles.

106 The work of Kinosita's group is described in Y. Arai *et al.*, 'Tying a Molecular Knot with Optical Tweezers', *Nature*, 399 (1999), 446.

107 The artificial molecular motor is described in N. Koumura, R. W. J. Zijlstra, R. A. van Delden, N. Harada, and B. L. Feringa, 'Light-Driven Monodirectional Molecular Motor', *Nature*, 401 (1999), 152.

108 Ross Kelly's rotary motor is described in T. R. Kelly, H. De Silva, and R. A. Silva, 'Unidirectional Rotary Motion in a Molecular System', *Nature*, 401 (1999), 150.

110 The field of synthetic molecular motors and machines is reviewed in V. Balzani, A. Credi, F. M. Raymo, and J. F. Stoddart, 'Artificial Molecular Machines', *Angewandte Chemie International Edition*, 39 (2000), 3348–91.

110 The work of Stan Leibler's group is described in F. J. Nédélec, T. Surrey, A. C. Maggs, and S. Leibler, 'Self-Organization of Microtubules and Motors', *Nature*, 389 (1997), 305.

111 Viola Vogel and colleagues reported their work in J. Dennis, J. Howard, and V. Vogel, 'Molecular Shuttles: Directed Motion of Microtubules along Nanoscale Kinesin Tracks', *Nanotechnology*, 10 (1999), 232.
The ATP synthase rotors made by Carlo Montemagno and colleagues are described in R. K. Soong *et al.*, 'Powering an Inorganic Nanodevice with a Biomolecular Motor', *Science*, 290 (2000), 1555–8.

6. Delivering the Message

115 The endocrine system and other systems of biomolecular communication are discussed in S. Aldridge, *Magic Molecules: How Drugs Work* (Cambridge: Cambridge University Press, 1998).

126 For an excellent discussion of neurotoxins and mind-altering drugs, see J. Mann, *Murder, Magic, and Medicine* (Oxford: Oxford University Press, 1992).

129 For an overview of supramolecular chemistry, see J.-M. Lehn, *Supramolecular Chemistry* (Weinheim: VCH, 1995). A less technical introduction is given in J.-M. Lehn and P. Ball, 'Supramolecular Chemistry', in N. Hall (ed.), *The New Chemistry* (Cambridge: Cambridge University Press, 2000).

130 The 'signal-transducing' synthetic molecular receptor is described in R. Krauss, H.-G. Weinig, M. Seydack, J. Bendig, and U. Koert, 'Molecular Signal Transduction through Conformational Transmission of a Perhydroanthracene Transducer', *Angewandte Chemie International Edition*, 39 (2000), 1835.

7. The Chemical Computer

133 John Hopfield speaks about biological existence theorems in the text of a talk entitled 'Biological information processing and molecular nanocomputing', given at the *Nature* conference on nanotechnology, Tokyo, January 1992.

137 The work of Leslie Orgel and others on replicating molecules is discussed in L. Orgel, 'Molecular Replication', *Nature*, 358 (1992), 203.

138 Günter von Kiedrowski reported his molecular replicators in G. von Kiedrowski, 'A Self-Replicating Hexadeoxynucleotide', *Angewandte Chemie International Edition*, 25 (1986), 932.

143 Nadrian Seeman's DNA cages are described in J. Chen and N. C. Seeman, 'Synthesis from DNA of a Molecule with the Connectivity of a Cube', *Nature*, 350 (1991), 631; Y. Zhang and N. C. Seeman, 'The Construction of a DNA Truncated Octahedron', *Journal of the American Chemical Society*, 116 (1994), 1661.

144 Chad Mirkin and colleagues reported their DNA-based method of nanoparticle assembly in C. A. Mirkin, R. L. Letsinger, R. C. Mucic, and J. J. Storhoff, 'A DNA-Based Method for Rationally Organizing Nanoparticles into Macroscopic Materials', *Nature*, 382 (1996), 607.

145 The peptides that recognize different semiconductor surfaces are described in S. R. Whaley, D. S. English, E. L. Hu, P. F. Barbara, and A. M. Belcher, 'Selection of Peptides with Semiconductor Binding Specificity for Directed Nanocrystal Assembly', *Nature*, 405 (2000), 665.

146 The molecule-based logic devices built by Jim Heath and co-workers are described in C. P. Collier *et al.*, 'Electronically

Configurable Molecular-Based Logic Gates', *Science*, 285 (1999), 391.

148 The rotaxane-based logic device is reported in A. Credi, V. Balzani, S. J. Langford, and J. F. Stoddart, 'Logic Operations at the Molecular Level: An XOR Gate Based on a Molecular Machine', *Journal of the American Chemical Society*, 119 (1997), 2679. The measurement of the electrical conductivities of single molecules is described in M. A. Reed, C. Zhou, C. J. Muller, T. P. Burgin, and J. M. Tour, 'Conductance of a Molecular Junction', *Science*, 278 (1997), 252; M. A. Reed and J. M. Tour, 'Computing with Molecules', *Scientific American* (June 2000), 86. Molecules that perform arithmetic are described in A. P. de Silva and N. D. McClenaghan, 'Proof-of-Principle of Molecular-Scale Arithmetic', *Journal of the American Chemical Society*, 122 (2000), 3965.

150 Leonard Adleman's proposal for DNA computing is described in L. M. Adleman, 'Molecular Computation of Solutions to Combinatorial Problems', *Science*, 266 (1994), 1021; L. M. Adleman, 'Computing with DNA', *Scientific American* (August 1994), 54.

Further reading

Atkins, P. W. *Atoms, Electrons, and Change* (New York: W. H. Freeman & Co., 1991).

—— *Molecules* (New York: W. H. Freeman & Co., 1991).

Ball, P., *Designing the Molecular World: Chemistry at the Frontier* (Princeton: Princeton University Press).

Emsley, J. *Molecules at an Exhibition* (Oxford: Oxford University Press, 1998).

Goodsell, D., *The Machinery of Life* (New York: Springer-Verlag, 1993).

—— *Our Molecular Nature: The Body's Motors, Machines and Messages* (New York: Springer-Verlag, 1996).

Hall, N. (ed.), *The Age of the Molecule* (London: Royal Society of Chemistry, 1999).

—— (ed.), *The New Chemistry* (Cambridge: Cambridge University Press, 2000).

Hoffmann, R., *The Same and Not the Same* (New York: Columbia University Press, 1995).

Index